Instilling Digital Competencies Through Educational Robotics

Ezio Fregnan
Catholic University of Milan, Italy & ESCP London, UK

Donatella Pinto
Independent Researcher, Italy

Giuseppe Scaratti
Catholic University of Milan, Italy & Bergamo University, Italy

A volume in the Advances in Educational Technologies and Instructional Design (AETID) Book Series

IGI Global
PUBLISHER of TIMELY KNOWLEDGE

Published in the United States of America by
IGI Global
Information Science Reference (an imprint of IGI Global)
701 E. Chocolate Avenue
Hershey PA, USA 17033
Tel: 717-533-8845
Fax: 717-533-8661
E-mail: cust@igi-global.com
Web site: http://www.igi-global.com

Library of Congress Cataloging-in-Publication Data

Names: Fregnan, Ezio, 1971- author. | Pinto, Donatella, 1961- author. | Scaratti, Giuseppe, 1957- author.
Title: Instilling digital competencies through educational robotics / by Ezio Fregnan, Donatella Pinto, Giuseppe Scaratti.
Description: Hershey PA : Information Science Reference, [2022] | Includes bibliographical references and index. | Summary: "This publication intend to provide some points for reflection and a first practical guidance for all those who are directly involved in the creation of new learning solutions, aimed at transferring value to citizens and workers of tomorrow"-- Provided by publisher.
Identifiers: LCCN 2021051156 (print) | LCCN 2021051157 (ebook) | ISBN 9781799886532 (Hardcover) | ISBN 9781799886549 (Paperback) | ISBN 9781799886556 (eBook)
Subjects: LCSH: Robotics in education. | Autonomous robots. | Science--Computer-assisted instruction. | Technology--Study and teaching. | Occupational training--Technological innovations. | Computer literacy. | Industry 4.0. | e.DO (Robot)
Classification: LCC LB1028.76 .F74 2022 (print) | LCC LB1028.76 (ebook) | DDC 371.33--dc23/eng/20211213
LC record available at https://lccn.loc.gov/2021051156
LC ebook record available at https://lccn.loc.gov/2021051157

This book is published in the IGI Global book series Advances in Educational Technologies and Instructional Design (AETID) (ISSN: 2326-8905; eISSN: 2326-8913)

British Cataloguing in Publication Data
A Cataloguing in Publication record for this book is available from the British Library.

For electronic access to this publication, please contact: eresources@igi-global.com.

Advances in Educational Technologies and Instructional Design (AETID) Book Series

ISSN:2326-8905
EISSN:2326-8913

Editor-in-Chief: Lawrence A. Tomei, Robert Morris University, USA

MISSION

Education has undergone, and continues to undergo, immense changes in the way it is enacted and distributed to both child and adult learners. In modern education, the traditional classroom learning experience has evolved to include technological resources and to provide online classroom opportunities to students of all ages regardless of their geographical locations. From distance education, Massive-Open-Online-Courses (MOOCs), and electronic tablets in the classroom, technology is now an integral part of learning and is also affecting the way educators communicate information to students.

The **Advances in Educational Technologies & Instructional Design (AETID) Book Series** explores new research and theories for facilitating learning and improving educational performance utilizing technological processes and resources. The series examines technologies that can be integrated into K-12 classrooms to improve skills and learning abilities in all subjects including STEM education and language learning. Additionally, it studies the emergence of fully online classrooms for young and adult learners alike, and the communication and accountability challenges that can arise. Trending topics that are covered include adaptive learning, game-based learning, virtual school environments, and social media effects. School administrators, educators, academicians, researchers, and students will find this series to be an excellent resource for the effective design and implementation of learning technologies in their classes.

COVERAGE

- Adaptive Learning
- Web 2.0 and Education
- Bring-Your-Own-Device
- Classroom Response Systems
- Online Media in Classrooms
- Higher Education Technologies
- E-Learning
- K-12 Educational Technologies
- Instructional Design
- Curriculum Development

IGI Global is currently accepting manuscripts for publication within this series. To submit a proposal for a volume in this series, please contact our Acquisition Editors at Acquisitions@igi-global.com or visit: http://www.igi-global.com/publish/.

The Advances in Educational Technologies and Instructional Design (AETID) Book Series (ISSN 2326-8905) is published by IGI Global, 701 E. Chocolate Avenue, Hershey, PA 17033-1240, USA, www.igi-global.com. This series is composed of titles available for purchase individually; each title is edited to be contextually exclusive from any other title within the series. For pricing and ordering information please visit http://www.igi-global.com/book-series/advances-educational-technologies-instructional-design/73678. Postmaster: Send all address changes to above address.

Titles in this Series

For a list of additional titles in this series, please visit:
https://www.igi-global.com/book-series/advances-educational-technologies-instructional-design/73678#titles

Design and Measurement Strategies for Meaningful Learning
José Luis Gómez Ramos (University of Castilla-La Mancha, Spain) and Isabel María Gómez-Barreto (University of Castilla-La Mancha, Spain)
Information Science Reference • copyright 2022 • 297pp • H/C (ISBN: 9781799891284) • US $195.00 (our price)

Technology-Supported Interventions for Students With Special Needs in the 21st Century
Xiongyi Liu (Cleveland State University, USA) and Patrick Wachira (Cleveland State University, USA)
Information Science Reference • copyright 2022 • 289pp • H/C (ISBN: 9781799888741) • US $195.00 (our price)

Cases on Practical Applications for Remote, Hybrid, and Hyflex Teaching
Valerie Harlow Shinas (Lesley University, USA) Chu N. Ly (Framingham State University, USA) and Sule Yilmaz Ozden (Sakarya Üniversitesi, Turkey)
Information Science Reference • copyright 2022 • 355pp • H/C (ISBN: 9781799891680) • US $195.00 (our price)

Affordances of Film for Literacy Instruction
Jason D. DeHart (Appalachian State University, USA)
Information Science Reference • copyright 2022 • 277pp • H/C (ISBN: 9781799891369) • US $195.00 (our price)

Applications of Machine Learning and Artificial Intelligence in Education
Seda Khadimally (North American University, USA)
Information Science Reference • copyright 2022 • 271pp • H/C (ISBN: 9781799877769) • US $195.00 (our price)

Handbook of Research on Promoting Global Citizenship Education
Jared Keengwe (University of North Dakota, USA)
Information Science Reference • copyright 2022 • 371pp • H/C (ISBN: 9781799895428) • US $265.00 (our price)

For an entire list of titles in this series, please visit:
https://www.igi-global.com/book-series/advances-educational-technologies-instructional-design/73678#titles

IGI Global
PUBLISHER of TIMELY KNOWLEDGE

701 East Chocolate Avenue, Hershey, PA 17033, USA
Tel: 717-533-8845 x100 • Fax: 717-533-8661
E-Mail: cust@igi-global.com • www.igi-global.com

Table of Contents

Foreword

Different types of intelligences can coexist in every individual, requiring a sufficiently diverse learning environment for these potentials to emerge. From visual-spatial skills to musical skills, from logical-mathematical skills to kinesics, from verbal-linguistic skills to interpersonal skills, each individual has within him or her a kaleidoscope of skills that need only be identified and trained.

The classical school system, unfortunately, tends to flatten and standardize all students to a single "standard intelligence", able to meet all the demands of the subjects of study. However, this inevitably leads to a levelling to an average value of the students' skills, which may not correspond to the real inclinations of the individual, or to the prevalence of one type of intelligence over another.

Such generalized learning environments to which are added classical teaching methodologies inevitably generate frustration, boredom, lowering of the motivation of students and therefore a drastic reduction in the drive to learn.

As far as the world of work is concerned, the White Paper on the future of Europe highlights how it is "probable that most children entering elementary school today will end up working in new types of jobs that do not yet exist"; this will require a refunding of the school system both in terms of the skills trained and the methodologies applied in the process.

The use of robotics turns out to be an extraordinary means in this scenario both in the development of professional skills and in the educational key, revolutionizing in a profound way the didactic methodology as in the STEM field, making the latter attractive and easier for young students that in this way will find stimulating educational contexts also favoring the overcoming of their weaknesses. Robots are machines that can perform complex operations if properly programmed and are objects that arouse the interest of young people and the desire to see them in action is an important motivational component.

The e.DO robots produced by Comau are a formidable ally for the teacher who thus becomes more of a facilitator of learning, encouraging the development of skills such as those of make decisions, problem solving, critical and creative thinking, effective communication, the ability to work in a group, which is part of the interpersonal ones; no less important is the ability to manage stress and emotions, also promoting awareness of their potential.

Therefore, robotics, if applied in a didactic/educational key, leads to countless advantages such as:

- greater student motivation;
- enhancement of soft skills;
- training of technical skills;
- development of computational thinking.

In this last aspect, the use of robotics encourages the decomposition of problems and their abstraction so that they can be solved through the design of algorithms. No less important is to be considered the inclusive dimension that this educational approach favors to the benefit of the most vulnerable students, and this becomes an important factor in combating early school leaving.

The volume accompanies us in a very concrete way, thanks to the scientific research approach adopted by the authors, in the fantastic world of robotics both in relation to the world of work and education to which the whole school system is called to an important innovative effort.

Paolo Cesana
Fondazione Clerici, Italy

Preface

We are in a scenario characterized by great changes (Lee et al., 2017): the Fourth Industrial Revolution and Digital Transformation are totally revolutionizing business models (Berman, 2012) and the approach with which companies approach the market (Bughin et al. 2019). The new technologies of the Internet of Things, Artificial Intelligence, and Robotics - to mention the main ones - are revolutionizing the way companies conceive, design, produce, distribute and sell their products and services (Fenech et al. 2019; Frank et al. 2019; Li et al, 2017). While this technological innovation has an exponential trend - so it's very fast - change in companies, organizations and people is proceeding more cautiously (Heawin, Power, 2018). These two curves with different accelerations create a gap: on the one hand the speed with which technology develops, on the other hand the slowness of organizations and human beings to adapt to it. This is why it is increasingly necessary to create a new work culture (Pinto et al, 2021; Heartl, Hess, 2017; Schein, 2000) that accompanies the development of new skills, consistent with the emerging social and organizational demands, capable not only to create new professions and new jobs, but also and above all to prepare and train people to a new way of working and living in a changing context (Pinto et al., 2021; Karimi, Walter, 2015).

This book seeks to connect the emergent transformations related to new ways of conceiving and realize productive and activity systems, with the Educational Robotics, represented as an intertwined world of people, practitioners, infrastructures, technologies, and a suitable investment for the creation of new shared value (Fregnan et al., 2020; Sapoundis, Alimisis, 2020; Alimisis 2013; Vial 2019).

The evolutionary process of Covid 19 has played a role of accelerator for potential social and organizational changes and has opened new challenges and trajectories for the future, soliciting also multiple dimensions and levels (Heavin, Power, 2018): economic, social, organizational, institutional, cultural, political.

The architecture of the book unfolds as follows: organizations and firms cannot avoid considering the social, environmental, and economic dimensions of their activity, facing the challenge of achieving common goods and generating shared values (Chapter 1). Hence the need to promote and develop innovative skills (Pinto et al., 2021; Van laar et al., 2017) enabling people (practitioners, workers, professionals) and particularly the new generations to cope with the unfolding radical evolutions, conveyed by the fourth industrial revolution and its transformations (Chapter 2). In this direction the Educational Robotics (Fregnan et al., 2020; Alimisis 2013) may represent an authentic learning environment (Herrington, 2014) for the acquisition of the 21st Century Skills (2008) (Chapters 3-4). The concrete experience of the e.Do project highlights how an investment in such a landscape can improve both work, professional and organizational cultures, and educational practices of teachers and students. The project's description and the main findings achieved, provide examples of social ergonomy (Talamo and Zucchermaglio, 2003; Zucchermaglio and Talamo 2000; 2003) of the use of robotics in education and for understanding the utility and efficacy of the e.Do (Chapters 4-7). The conclusions (Chapter 8) underpin how the described experience can be adopted both by a corporate audience as a demonstration of creating shared value, and by educational robotic experts as an example of a robotic project deployed on a wide scale.

THE RESEARCH

The scientific research described in this book questions the value of Educational Robotics initiatives (Fregnan et al., 2019; Saavedra 2012) in developing what we can generically define as 21st century skills and capabilities (Pinto et al., 2021; Warner, Wäger, 2019). Educational Robotics seems to be a powerful tool to develop new skills, consistent and appropriate to the new scenario in which we are immersed, but the current vision of Educational Robotics is limited and often there seems to be not enough awareness about the multiple values that it can generate even in fields far from pure robotics, and it is not encouraged and stimulated outside of specific niches. With this research authors therefore set out to give visibility to the great and multiple values that Educational Robotics initiatives can offer to all those involved as participants, as well as to those who design and implement the initiatives and other stakeholders engaged in improving training and educational solutions for the future of work.

A robust Educational Robotics project has been identified, the one experienced by the Comau Academy with the proprietary robot e.DO and the *e.DO Experience* (Fregnan et al., 2019), and authors have set the goal of recounting the richness of it. *e. DO Experience*, in fact, is an interesting project not only for Comau - which is a technological, international and innovative company - but also because it is a new idea, driven by a key thought: "let's create a robot called e.DO, let's make it become the ambassador of Digital Transformation and of the Fourth Industrial Revolution and let's use it to spread a new work culture and new skills in an innovative way".

The challenge underlying this initiative is not only to create robotics skills, but also to spread work culture, STEM (Aris, Orcos, 2019), soft skills, involving and exciting young people (Benitti, 2012). In this approach lies the innovative dimension of the initiative and its connection with the theme of value creation (Porter, Kramer, 2011): the e.DO robot has allowed to break some technological barriers and to expand the space of action towards issues broader than the specific one of robotics training: "a robot to spread values" seems a contradiction in terms, and on this apparent contradiction authors have focused to analyze in a scientific way the values generated by Educational Robotics initiatives.

Problem, Aim, and Goals

Problem

The Digital Transformation is a revolutionary phenomenon affecting society, economy, culture, and work. This new scenario also has an impact on organizations and people: we are observing brand-new trends and needs in terms of society and companies' evolution and the consequent demand for new skills for their employees. Educational robotics seems to be **a powerful tool for developing new skills**, coherent and appropriate to the new scenario in which we are immersed. Meanwhile, **the current vision on educational robotics is limited**, as there is no awareness of the **multiple value it can bring even in fields far from pure robotics**, and it is not encouraged and stimulated beyond specific niches.

Aim

The research aims to give **visibility to the great and multiple values that educational robotics initiatives can offer** to all the people involved as participants, as well as to those who design and implement the initiatives, and to other stakeholders committed in improving training and educational solutions for the future of work.

Goals

1. **Describing** the Digital Transformation scenario, the society and companies' evolution and the request for new skills
2. **Identifying** the specific values of educational robotics initiatives, also describing an innovative educational robotics project: the Comau *e.DO Experience*
3. **Verifying and exploring** the values generated by the Comau *e.DO Experience* and identifying, listing and describing the main indicators for each value
4. **Measuring** the weight of each value and its respective indicators in the perception of those who designed and delivered the Comau *e.DO Experience* educational robotics project
5. **Exploring** perceptions and representations of participants in the *Comau e.DO Experience* educational robotics project

Stakeholders

The research is focused on a topic of great relevance and impact for anyone involved in training: **educational robotics as a lever to spread digital skills**. In particular, authors believe the research can provide food for thought and suggestions useful for:

- *People who design and implement educational robotics initiatives*

The evidences that emerge from the research are concrete planning and operational ideas to give life to training initiatives consistent with the needs of people who face the world of work of the future and who must develop different types of skills, by means of innovative and effective learning methods.

- *People who participate in educational robotics initiatives*

The evidences that emerge from the research are a useful tool to orient yourself among the various training and educational proposals. Being aware of the value offered by "well-done" educational robotics initiatives allows people to choose the most appropriate solutions with greater awareness and intentionality.

- *Other stakeholders committed to improve training and educational solutions for the future of work*

The evidences emerging from this research are also interesting for all the stakeholders, both internal (who lives the organization on a daily basis) and external (who is directly or indirectly involved in training and educational initiatives). For both groups, a more precise awareness of the value of educational robotics initiatives can represent a central element for their decision-making processes.

Methodology

According to the research objectives, authors decided to adopt a **mixed methodology** to collect and process all useful data to achieve them. More in detail, in Table 1 is described the approach in terms of **research methodology**, **tools** and **sample**.

The research is divided into 3 phases, consistent with the research objectives identified.

1. *The first phase* aimed to investigate the specific values of educational robotics initiatives. By means of desk-based research, authors investigated the scenario of Digital Transformation, the evolution that is taking place in organizations and the consequent need to provide new skills to the people who are or who will be part of it.

They then dedicated themselves to explore educational robotics initiatives, in order to verify what kind of contribution they can offer to these new needs, from a CSV perspective. In this case, the desk-based research involved both the scientific literature and the analysis of a real and concrete project: the Comau *e.DO Experience*. The results of the desk-based research have provided the authors with the theoretical foundations to elaborate the next steps of their work, and in particular to start identifying what are in their opinion the areas, the levels and the indicators of values generated by educational robotics initiatives. This allowed them to create the first tool for a field research, the interview outline for interviewer, aimed at verifying and exploring all the elements.

Table 1. Research Methodology

GOAL	METHODOLOGY	TOOL	SAMPLE
1. Describing the Digital Transformation scenario, the society and companies' evolution and the request for new skills	Desk-based research	/	/
2. Identifying the specific values of educational robotics initiatives, also describing an innovative educational robotics project: the Comau *e.DO Experience*	Desk-based research	/	/
3. Verifying and exploring the values generated by the Comau *e.DO Experience* and identifying, listing and describing the main indicators for each value	Qualitative interviews	Interview outline for interviewer	14 interviews: 7 Comau Managers 7 Comau Partners
4. Measuring the weight of each value and its respective indicators in the perception of those who designed and delivered the Comau *e.DO Experience* educational robotics project	Survey	Survey Monkey SPSS statistical software	124 surveys: • 45 Comau people • 79 Comau Partners
5. Exploring perceptions and representations of participants in the Comau *e.DO Experience* educational robotics project	Semiotic analysis	Focus Group	• 20 primary school students • 20 first grade secondary school students • 20 second grade secondary school students

2. *The second phase* had an exploratory purpose and aimed to verify whether the value indicators for educational robotics initiatives are really perceived as such by those who have actually implemented initiatives of this type - in this case the Comau managers involved in the *e.DO Experience* - and by external stakeholders - in this case some of the partners belonging to the Comau network. Therefore, the second phase included 14 semi-structured interviews with Comau managers and partners. The interviews allowed the authors to rediscover and deep-dive the presence of the elements hypothesized. Thanks to this, they have been able to create the quantitative survey for the third phase of the research.
3. *The third phase* was useful to size the different elements of values validated together with the interviewees, examining the weight of each

Table 2. Phases of the Research Work

PHASE 1 (January-March 2021)	PHASE 2 (April-June 2021)	PHASE 3 (July-September 2021)
Desk-based research	Semi-structured qualitative interviews administration	Survey administration
Interview outline creation	Survey creation	Focus Group semiotic analysis

of them on the perception of shared value. Authors administrated a quantitative survey to 124 people (45 Comau people and 79 partners) and they collected the data to measure all the items. In parallel, in order to grasp the meanings attributed to the *e.DO Experience,* the use of focus groups was adopted, as functional to a semiotic, narrative and conversational reconnaissance of the experienced training situation. A sample of 20 primary school students, 20 first grade secondary school students and 20 second grade secondary school students was randomly selected from participants in the workshop, foreshadowing the activation of 3 focus groups for each school level.

The research work **lasted 9 months**, from January 2021 to September 2021. Figure 1 shows the **summary scheme** of the entire research.

Figure 1. Summary scheme of the research

GOAL	METHODOLOGY	TOOL & SAMPLE	MAIN FINDINGS	CONCLUSIONS
1. Describing the Digital Transformation scenario, the society and companies' evolution and the request for new skills	Desk-based research	/	• Description of the main transformations that occurred with DT • Description of the impact of DT on companies' evolution • Identification of the skills necessary for organizations and people to cope with this transformative scenario	Within the DT scenario, value creation occurs through the development of skills that enable people to live and work in a 4.0 world: STEM Skills and Capabilities
2. Identifying the specific values of educational robotics initiatives, also describing an innovative educational robotics project: the Comau *e.DO Experience*	Desk-based research	/	• Description of main implications in using robots for education • Understanding of the main differences between Robotic Education and Educational Robotics • First identification of the areas of values, the levels of value and the indicators for each area of value generated by educational robotics initiatives such as the *e.DO Experience* • Elaboration of the summary synoptic • Development of the interview outline	In terms of value creation, educational robotics initiatives serve several functions, particularly related to their effects on the participants: - Skill Creation - Value for New Citizenship - Empowerment - Inspiration & Orienteering The Comau *e.DO Experience* can also direct the creation of value at a company and territorial level in terms of commons and generativity - Value for Company - Value for Territory Authors will verify the hypothesis by means of a qualitative interview and a survey dedicated to Comau managers and partners involved in the *e.DO Experience* predoject.

ORGANIZATION OF THE BOOK

The book is organized in eight chapters that trace the work done to meet the identified research objectives. A brief description of each of the chapters follows.

Chapter 1 introduces the research project as a whole and investigates the issue of the value that today's organizations can contribute to generate internally and externally, through an overview of how the concept of value itself has evolved in organizations (Porter, Kramer, 2011, 2019; Assiouras et al., 2019; Crane et al., 2019). The analysis carried out identifies commons (Ostrom, 1990; De Angelis, 2003; 2006;2017; De Angelis, Harvie, 2013; Centemeri, 2018) and generativity (Stiegler 2010; 2010a; Magatti, Giaccardi 2017; Gherardi, Magatti, 2014; Bradley; Marcia, 1998) as the concepts from which to define a transversal value matrix, useful as a general reference framework.

Chapter 2 investigates the Digital Transformation scenario, the evolution of society and businesses (Vial, 2019; Berman 2012) and the emerging demand for new skills (Pinto et al., 2021).

Chapter 3 and Chapter 4 are dedicated to investigating the specific values of Educational Robotics initiatives by cross-referencing the evidence emerging from the scientific literature (Alimisis 2013; Alves-Oliveira et al., 2019; Belpeame et al., 2018; Benitti; 2012) and from the analysis of Comau's *e.DO Experience* project (Fregnan et al., 2020). Thanks to this research and analysis activity, it was possible to proceed to an initial identification of value levels, value areas and indicators generated by Educational Robotics initiatives such as the *e.DO Experience*.

Chapter 5 reports some preliminary conclusions from the first phase of research. The evidence collected, in fact, allowed to identify the specific values of Educational Robotics initiatives in terms of commons and generativity and - thanks to the precise analysis of the *e.DO Experience* initiative - allowed to map all the value elements generated by Educational Robotics initiatives: levels, areas, indicators.

Chapter 6 outlines as a starting hypothesis that the Educational Robotics initiatives undertaken by Comau through the *e.DO Experience* project have enabled the generation of shared value among Participants, Company, and Territory. The assumption is that the many promoters, whether Comau managers or third-party company partners, were motivated by shared goals and interests but, at the same time, that their perceptions and representations of the different activities and the value they generated may have peculiar and unique elements. In order to capture the variety of different perspectives, so as to return a complete and in-depth picture of the different levels of value generated for Participants, Company and Territory, authors have therefore prepared a qualitative interview with 7 managers and 7 Comau partners involved in the *e.DO Experienc*e project.

Chapter 7 aims - in the light of what emerged from the qualitative interviews - to measure the weight of each value and the respective indicators that emerged, investigating the perception of those who designed and implemented the Educational Robotics project *e.DO Experience*. To this end, a Survey was prepared involving 124 people (45 Comau employees and 79 Comau partners) who were asked to express their perception of the usefulness of the *e.DO Experience* initiative to generate value in Participants, Company and Territory.

Finally, Chapter 8 turns its attention to the point of view of participants in Educational Robotics initiatives. To capture the meanings attributed to the *e.DO Experience*, the use of focus groups was adopted, functional to a semiotic, narrative, and conversational reconnaissance (Greimas 2000; Ricoeur, Greimas 2000) of the training situation experienced, soliciting a process of social construction of the value attributed to what happened during the initiative.

Authors hope that the reading of this research, which has brought out an important hidden value related to the *e.DO Experience* educational robotics initiative (Fregnan et al., 2019), may represent a stimulus and a model from which to take inspiration for future initiatives. The diffusion of this kind of Educational Robotics projects, which are based on the creation of new ecosystems and on a new educational perimeter in which young people, universities, companies, and institutional bodies work in a synergic way (Pinto et al., 2018), can be an important contribution for the construction of a more human future, characterized by a strong technological diffusion, but sustainable from a social, environmental, and economic point of view.

REFERENCES

Alimisis, D. (2013). Educational robotics: Open questions and new challenges. *Themes in Science and Technology Education*, *6*(1), 63–71.

Alimisis, D. (2019). Teacher training in educational robotics: The ROBOESL Project paradigm. *Technology. Knowledge and Learning*, *24*(2), 279–290. doi:10.100710758-018-9357-0

Alves-Oliveira, P., Sequeira, P., Melo, F. S., Castellano, G., & Paiva, A. (2019). Empathic robot for group learning: A field study. *ACM Transactions on Human-Robot Interaction*, *8*(1), 1–34. doi:10.1145/3300188

Arís, N., & Orcos, L. (2019). Educational robotics in the stage of secondary education: Empirical study on motivation and STEM skills. *Education Sciences*, *9*(2), 73. doi:10.3390/educsci9020073

Assiouras, I., Skourtis, G., Giannopoulos, A., Buhalis, D., & Koniordos, M. (2019). Value co-creation and customer citizenship behavior. *Annals of Tourism Research*, *78*, 102742. doi:10.1016/j.annals.2019.102742

Belpaeme, T., Kennedy, J., Ramachandran, A., Scassellati, B., & Tanaka, F. (2018). Social robots for education: A review. *Science Robotics*, *3*(21), 21. doi:10.1126cirobotics.aat5954 PMID:33141719

Benitti, F. B. V. (2012). Exploring the educational potential of robotics in schools: A systematic review. *Computers & Education*, *58*(3), 978–988. doi:10.1016/j.compedu.2011.10.006

Berman, S. J. (2012). Digital transformation: Opportunities to create new business models. *Strategy and Leadership*, *40*(2), 16–24. doi:10.1108/10878571211209314

Bradley, C. L., & Marcia, J. E. (1998). Generativity-stagnation: A five-category model. *Journal of Personality*, *66*(1), 39–64. doi:10.1111/1467-6494.00002

Bughin, J., Deakin, J., & O'Beirne, B. (2019). Digital transformation: Improving the odds of success. *The McKinsey Quarterly*.

Centemeri, L. (2018). Commons and the new environmentalism of everyday life. Alternative value practices and multispecies commoning in the permaculture movement. *Rassegna Italiana di Sociologia*, *59*(2), 289–314.

Crane, A., Palazzo, G., Spence, L. J., & Matten, D. (2014). Contesting the value of "creating shared value". *California Management Review*, *56*(2), 130–153. doi:10.1525/cmr.2014.56.2.130

De Angelis, M. (2003). Reflections on alternatives, commons and communities. *The Commoner*, *6*(Winter), 1–14.

De Angelis, M. (2006). The World Social Forum: Challenging empires. *Development*, *49*(2), 125–128. doi:10.1057/palgrave.development.1100264

De Angelis, M. (2017). *Omnia sunt communia: On the commons and the transformation to postcapitalism*. Bloomsbury Publishing. doi:10.5040/9781350221611

De Angelis, M., & Harvie, D. (2013). *The commons*. Routledge.

Fenech, R., Baguant, P., & Ivanov, D. (2019). The changing role of human resource management in an era of Digital Transformation. *Journal of Management Information and Decision Sciences*, *22*(2), 1–10.

Frank, A. G., Mendes, G. H., Ayala, N. F., & Ghezzi, A. (2019). Servitization and Industry 4.0 convergence in the digital transformation of product firms: A business model innovation perspective. *Technological Forecasting and Social Change*, *141*, 341–351. doi:10.1016/j.techfore.2019.01.014

Fregnan, E., Ivaldi, S., & Scaratti, G. (2020). HRM 4.0 and New Managerial Competences Profile: The COMAU Case. *Frontiers in Psychology*, *11*, 11. doi:10.3389/fpsyg.2020.578251 PMID:33329228

Gherardi, L., & Magatti, M. (2014). *Una nuova prosperità: Quattro vie per una crescita integrale*. Feltrinelli Editore.

Greimas, A. J. (2000). *Semantica strutturale: ricerca di metodo* (Vol. 2). Meltemi Editore srl.

Hartl, E., & Hess, T. (2017). *The role of cultural values for digital transformation: Insights from a Delphi study*. Academic Press.

Heavin, C., & Power, D. J. (2018). Challenges for digital transformation–towards a conceptual decision support guide for managers. *Journal of Decision Systems, 27*(sup1), 38-45.

Karimi, J., & Walter, Z. (2015). The role of dynamic capabilities in responding to digital disruption: A factor-based study of the newspaper industry. *Journal of Management Information Systems*, *32*(1), 39–81. doi:10.1080/07421222.2015.1029380

Lee, M. X., Lee, Y. C., & Chou, C. J. (2017). Essential implications of the digital transformation in industry 4.0. *Journal of Scientific and Industrial Research*, *76*, 465–467.

Li, G., Hou, Y., & Wu, A. (2017). Fourth Industrial Revolution: Technological drivers, impacts and coping methods. *Chinese Geographical Science*, *27*(4), 626–637. doi:10.100711769-017-0890-x

Magatti, M., & Giaccardi, C. (2014). *Generativi di tutto il mondo unitevi! Manifesto per la società dei liberi*. Feltrinelli.

Ostrom, E. (1990). *Governing the commons: The evolution of institutions for collective action*. Cambridge university press. doi:10.1017/CBO9780511807763

Pinto, D., Scaratti, G., & Fregnan, E. (2018). *Giovani, Università e Azienda: il nuovo perimetro formativo*. Pearson.

Pinto, D., Scaratti, G., & Fregnan, E. (2021). *Learning Agility 4.0: ecosistemi e trasformazione culturale*. Pearson.

Porter, M., & Kramer, M. R. (2011). Creare valore condiviso. *Harvard Business Review Italia*, *1*(2), 68–84.

Porter, M. E., & Kramer, M. R. (2019). Creating shared value. In *Managing sustainable business* (pp. 323–346). Springer. doi:10.1007/978-94-024-1144-7_16

Ricoeur, P., & Greimas, A. J. (2000). *Tra semiotica ed ermeneutica* (Vol. 4). Meltemi Editore srl.

Saavedra, A. R., & Opfer, V. D. (2012). Learning 21st-century skills requires 21st-century teaching. *Phi Delta Kappan*, *94*(2), 8–13. doi:10.1177/003172171209400203

Sapounidis, T., & Alimisis, D. (2020). *Educational robotics for STEM: A review of technologies and some educational considerations. In Science and Mathematics Education for 21st Century Citizens: Challenges and Ways Forward*. Nova Science Publishers.

Schein, E. H. (1999). The Corporate culture survival guide. Jossey-Bass.

Stiegler, B. (2010). For a new critique of political economy. *Polity*.

Stiegler, B. (2010a). *Taking care of youth and the generations*. Stanford University Press.

Talamo, A., & Zucchermaglio, C. (2003). *Inter@zioni. Gruppi e tecnologie*. Carocci.

Van Laar, E., Van Deursen, A. J., Van Dijk, J. A., & De Haan, J. (2017). The relation between 21st-century skills and digital skills: A systematic literature review. *Computers in Human Behavior*, *72*, 577–588. doi:10.1016/j.chb.2017.03.010

Vial, G. (2019). Understanding digital transformation: A review and a research agenda. *The Journal of Strategic Information Systems*, *28*(2), 118–144. doi:10.1016/j.jsis.2019.01.003

Warner, K. S., & Wäger, M. (2019). Building dynamic capabilities for digital transformation: An ongoing process of strategic renewal. *Long Range Planning*, *52*(3), 326–349.

Zucchermaglio, C., & Talamo, A. (2000). The Social Construction of Work Times: Negotiated time and expected time. *Time & Society*, *9*(2-3), 205–222.

Zucchermaglio, C., & Talamo, A. (2003). The development of a virtual community of practices using electronic mail and communicative genres. *Journal of Business and Technical Communication*, *17*(3), 259–284.

Acknowledgment

Comau People

Duilio Amico
Nicoletta Beretta
Malcolm Bisio
Giuseppe Costabile
Giuseppe Daqua
Serena Dal Piaz
Francesco Faccio
Lucia Marchesi
Stefano Pesce
Marta Serrano
Luca Vinci
Marina Zussino

Comau Partner

Matteo Bonanno (Pearson Italia)
Paolo Cesana (Fondazione Clerici)
Roberto Dentale (Asse4)
Marco Gioannini (Fondazione Agnelli)
Matteo Pisanu (Digital Innovation Hub)
Manuel Tonolini (Fondazione Dalmine)
Diego Teloni (MakerDojo)

Chapter 2 was written with the contribution of Silvia Ivaldi, University of Bergamo.

Chapter 3 was written with the contribution of Leonardo Ciocca, Università Cattolica del Sacro Cuore di Milano and Francesca Grignolio, Consultant for Comau Education Business.

Chapter 6 was written with the contribution of Leonardo Ciocca, Università Cattolica del Sacro Cuore di Milano and Francesca Grignolio, Consultant for Comau Education Business.

Chapter 7 was written with the contribution of Leonardo Ciocca, Università Cattolica del Sacro Cuore di Milano and Francesca Grignolio, Consultant for Comau Education Business.

Chapter 8 was written with the contribution of Silvia Ivaldi, University of Bergamo.

Duilio Amico
Nicoletta Beretta
Malcolm Bisio
Giuseppe Costabile
Giuseppe Daqua
Serena Dal Piaz
Francesco Faccio
Lucia Marchesi
Stefano Pesce
Marta Serrano
Luca Vinci
Marina Zussino
Matteo Bonanno (Pearson Italia)
Paolo Cesana (Fondazione Clerici)
Roberto Dentale (Asse4)
Marco Gioannini (Fondazione Agnelli)
Matteo Pisanu (Digital Innovation Hub)
Manuel Tonolini (Fondazione Dalmine)
Diego Teloni (MakerDojo)

Chapter 1
The Concept of Value

ABSTRACT

The actions of a company can generate value on different levels: economic, environmental, and social. Recent looks at the concept of value have outlined significant implications, able to go beyond a concept, albeit important but reductive, in terms of maximizing profits and earnings for investors. Starting from the theory of shared value of Porter and Kramer and passing through critical management studies, the authors have come to consider value in a perspective oriented to the exchange and continuous growth of a company and of all the different territorial and social stakeholders involved, in a responsible, generative sense and aimed at the common good.

INTRODUCTION

Organizations, through the exercise of their activities, create value and its maximization can be interpreted as the ultimate goal of the enterprise (Dembek, Singh, Bhakoo, 2015).

But in what terms can this value be described?

The multitude of facets that characterize it puts this object of study at the center of interdisciplinary attention, including as a focus the contributions of economic, environmental and social sciences. (Dembek, Prakash, Bhakoo, 2016).

A choice of perspective is required within the literature, due to the large and wide availability of perspectives; this research aims to present the concept of value with particular reference to its social dimension, considering it as

DOI: 10.4018/978-1-7998-8653-2.ch001

a sharing process involving the organization and its various stakeholders, in which economic, environmental and social interests are integrated into the core business (Porter, Kramer, 2011, 2019).

In this chapter will be presented an **overview of how the concept of value has evolved** from its earliest attributions to the definition of Creating Shared Value (CSV), a concept in which organizations play a central role in creating and developing social value shared with its stakeholders (Porter, Kramer 2011, 2019).

CREATING SHARED VALUE

From Value to Social Value

The first approaches to the concept of value tend to highlight a clear trichotomy that can be tracked back to the main scientific fields that investigate it. It is possible to identify three distinct levels of the concept of value: economic, environmental and social (Amhoj, 2019). For an organization, creating economic value means to generate enough profit to ensure an adequate return on the investment (Dembek, Prakash, Bhakoo, 2016). On the other hand, the creation of environmental and social value means an active participation in organizational challenges by recognizing not only an economic, but also a social responsibility of the company, aimed to preserve the environment and to promote welfare for both community and workers (Dembek, Prakash, Bhakoo, 2016; Porter, Kramer, 2011; 2019). From the first perspectives, these three dimensions are considered independent. The most significant implications, in terms of profit maximization and generation of competitive edge, are promoted and enhanced (Barney, 1991).

Although these elements are essential for an organization, just focusing on them could represent a risk for the complete understanding of the object of study. More recent studies, however, have started to consider the **social, environmental, and economic dimensions as interconnected elements** included in what Porter and Kramer (2011; 2019) define as **shared value**.

Creating Shared Value

Creating shared value means creating economic value in a way that also produces environmental and social value, by incorporating environmental and social challenges into the company's business model (Porter, Kramer, 2011; 2019).

The greatest innovation represented by the concept of shared value is that **strategies and initiatives in support of society become part of the business model**. This is a new way to lead the organization to a sustainable long-term profit and a mutual exchange of prosperity with community and society as a whole. (Porter, Kramer, 2011; 2019) There are three ways to create shared value:

1. Products and markets need to be redefined addressing social issues such as health, supporting quality nutrition or reducing environmental damage;
2. To re-plan productivity by paying attention to social and environmental matters such as water waste, excessive waste generation, as well as working and community conditions;
3. To develop and sustain local clusters supporting organizational productivity, innovation, and competitiveness (Porter, Kramer, 2011; 2019*)*.

A great interest is growing for the concept of CSV among scholars and, above all, a larger number of companies. This achievement is accomplished thanks to authors' ability to represent the social responsibility of organizations, connoting **social challenges as strategic opportunities for business**, rather than costs imposed from the outside. (Crane et al., 2014).

Within this perspective, the organization and its Human Resources Management (HRM) functions find themselves at the center of tensions generated by multiple and not always aligned interests (Galuppo et al., 2014). The **multitude of stakeholders involved in the CSV processes engendered complexity**: for a better understanding, an overview on the different involved stakeholders will be presented in the next section.

Multistakeholdership and CSV

If early perspectives about the concept of value are addressed only to shareholders with at short-term economic advantage (Stout, 2019), more recent studies aim to extend the engagement to all of the firm's stakeholders, whether internal or external. (Hillman, Keim, 2001). The **internal stakeholders** of an organization can be considered the individual workers, managers and entrepreneurs who live the organization on a daily basis; on the other hand,

external stakeholders are all the actors directly or indirectly involved in the business processes, such as customers, suppliers, the territory in which the organization is located in, the communities to which the company belongs and other kind of partners (Galuppo et al, 2014). Porter and Kramer (2011; 2019) emphasize the dimensions of corporate responsibility in the active and integrated involvement of an organization's external and internal stakeholders. According to Husted and Allen (2000), the concept of CSV refers to the possibility of creating economic value for the firm and its stakeholders in conjunction with environmental and social value. CSV stands as a new look, able to outline the strategies that a company can adopt in its business, in terms of tangible, intangible and human resource management, energy use, responsible logistics organization and rethinking of the employee productivity, as well as its relationship with the community (Porter, Kramer, 2011; Husted, Allen, 2011; Ketchen, Hult, 2007). A company's value chain inevitably affects and is affected by numerous social issues, that include: the exploitation of natural and human resources, the health and welfare assessment of the local communities in which the company is located, the conditions, safety, rights, and treatment of the employees in the workplace (Porter, Kramer, 2011; 2019; Ketchen, Hult, 2007) It is therefore important to take into consideration that social issues that are not taken up by corporate policies can create ethical and reputational problems in the corporate value chain (De Los Reyes, Scholz, 2019).

The CSV Theory has been very useful in recognizing the vast panorama of stakeholders with whom the company has relationships, laying the roots for a new vision that transcends the mere importance in terms of profit, marketing and reputation improvement, capable of **considering value in close connection with issues of HRM, sustainable development, corporate reputation and corporate social responsibility** (Crane et al. 2014). In this regard, a critical perspective to CSV will be presented in the next paragraph.

CSV: A Critical Perspective

Adopting a more critical approach, authors want to question whether the commitment of organizations to CSV programs is directed by marketing policies (to improve their image and promote the loyalty of clients and suppliers), or whether it represents a true commitment to business ethics (Matinheikki, Rajala, Peltokorp, 2017). The role that corporate reputation has in organizational results (the World Economic Forum attests that it is worth 25% of a company's market value), as well as the rising socialwashing and greenwashing phenomenon, reveal that the implications in terms of marketing, which have to do with CSV, seem to justify the importance of a critical perspective. What is at stake is the way on how CSV could be proposed: the risk is to improve corporate reputation and CSR just for economic purposes (Prasad, Holzinger, 2013). Obviously, the intention to exploit the profitability linked to CSV is more than legitimate, but if the sharing of that value is not supported by a real commitment to business ethics, the result could be dangerous for the company reputation itself. Reputation, CSR, attractiveness and customer loyalty, which can be considered as the major drivers of extra profit associated with the development of a CSV policy, represent an opportunity to **lead companies towards the introduction of social challenges into their business**, which is the cornerstone of Porter and Kramer's (2011; 2019) shared value conception. It is indeed true, and it also justifies the enormous success that the concept of shared value has produced, that in order to achieve a sustainable and social commitment of organizations, the economy fundamentals and the rule of convenience have to be followed, and that goes beyond philanthropy (Matinheikki, Rajala, Peltokorp, 2017).

It remains a critical question to determine **how engagement in social value creation can represent a true and ethical commitment** for the organization. In this regard, Prasad and Holzinger (2013) identify two key conditions: the first is that the social commitment includes an open and recurrent dialogue not only with the stakeholders most involved in the process, but also with the more emarginated classes of stakeholders. The second, and perhaps the most significant condition, is to promote a change in the organizational culture driven by HRM to integrate the values of social and environmental sustainability into the "organizational machine." (De Los Reyes, Scholz, 2019) At this point, to open a new space for HRM within organizational landscapes, linking this function to the management of CSV processes through an authentic CSR, it is possible to refer to Critical Management Studies (CMS) approach (Alvesson and Willmott, 1992, 1996; Alvesson and Deetz, 2000; Alvesson and Kärreman, 2007; Alvesson et al., 2008). Challenges and connections entail the possibility of reviewing the notion of HRM itself, examining new research perspectives and lines of interpretation following a Critical Management Studies approach, thus developing a more contextualized view in conceiving HRM and CSV, a more expansive consideration of stakeholders, as well as a longer-term perspective. In this regard, CSV is seen as **a social practice embedded in specific situated contexts**. Such conception enables the involvement of multiple rationalities, related to both internal and external stakeholders, achieving forms of reconstructive reflexivity concerning the interconnection between HRM and the innovative generation of social value through an authentic corporate responsibility (Fregnan, Ivaldi, Scaratti, 2020).

NEW PERSPECTIVES: COMMONS AND GENERATIVITY[1]

In this section, the purpose is not to discuss directly about what organizations are doing to share value, but rather to focus on new ways of thinking about markets and society related to the concept of CSV (Butera, 2020). The first conceptions identified the exploitation of resources as the primary way to create economic value, but we are now witnessing a reversal: the key to the creation of value **lies in the protection and promotion of resources, in their involvement in a process of exchange and sharing governed by moral principles of social justice, sustainable growth, community, ethics, prosperity and genuine responsibility** (Porter, Kramer, 2011; 2019).

Authors will discuss **commons and generativity**, concepts from which they will take their cue to define a **value matrix** to be used as a reference throughout the text.

Commons

The commons are common goods and services, both tangible and intangible, that originally are neither the property of public authorities such as the city or state, nor the private property of individuals or businesses (De Angelis 2006; 2017). The author David Bollier (2010), starting from the work of the American Nobel Price Elinor Ostrom (1990), based the definition of commons on three foundational elements (Centemeri, 2018):

- The commons are a common property, managed, preserved and created, by a community;
- The commons are not simply something that belongs and is accessible to all, but something that depends on human decisions and value creation activities. The commons are not a natural phoenomenon
- This common property is managed in autonomu and in relation with community rules and regulations.

Commons are circumscribed by the following criteria: 1) commons are an activity carried out by people 2) commons are a collaborative object,, 3) commons are a form of ownership and management (Bollier, 2010). To understand the commons as a socio system structured by alternative value practices (Centemeri 2018), we will focus on the proposal of Massimo De Angelis (2006; 2017): De Angelis' framework is one of the most developed and accurate examples of a critical theory of the commons. Common goods, namely the *commonwealth*, are just one part of the commons and they are related to other elements: the *commoners*, which represent the social subject and the social activity of doing in common, or *commoning* (De Angelis, 2017; Centemeri 2018), towards the maintenance and reproduction of the socio-ecological system and its components. As Quoting De Angelis (2017), Centemeri (2018) suggest that commonwealth is wealth «because a plurality claims it» (De Angelis 2017, 136) and this plurality is the community of commoners, «with their affective and social relations» (De Angelis, 124). The commoners' community is based on a certain quality of relations between people and towards goods and resources that constitute their commonwealth. Commons introduce a new perspective to understand the concept of value, to fulfill social needs, to manage social production and to support the CSV. Commons stands in communities, by means of social networks of mutual support, solidarity, and practices of human exchange (Centemeri, 2018).

The generation of communities on the basis of commons is related to the interactions among individuals and communities as human interactions that give rise to many different kinds of exchanges. Communities are everywhere a non-competitive form of relation among human beings exists (De Angelis 2016; De Angelis, Harvie 2014). In common parlance, however, we refer to the word *community* to define a group of people who share something, and the content of this sharing will shape the specific nature of a certain community (De Angelis, 2017). Commons practices, including those based on generative economies, are based on the belief that the creation of shared goods and services occurs in a context of mutual interest (Pais, Provasi, 2015). The commons put the basis for a new way to conceive organization as generative agents, acting to preserve the commons and to develop communities (Centemeri, 2018).

At stake is the need to **think and promote market forms capable of preserving the commons**, rather than exploiting them. The aim is not only **value generation**, but also its **reinvestment in the commons and the communities**. Finally, the commons are a clear challenge to traditional forms of market and civil society (Ostrom et al., 1994; De Angelis, 2003; 2016); at an organizational level they bring **new forms of coordination and management**, which are much more based on informal participation, voluntary action and horizontal management practices, often critical of exclusive forms of professionalization and managerialism (Centemeri, 2018).

Generativity

Porter and Kramer's (2011; 2019) contribution shifts the focus to the need of re-connecting economy and society, proposing new ways to interpret the markets on a generativity base (Magatti, Giaccardi 2014): "If consumption consumes, it is difficult to think that sustainable development can be planned on it. The center of gravity of politics, society and the economy must be changed: it cannot be consumption. Rather, the new center of gravity must be sought in the human capacity to generate value. It is time to take a broader view of value creation, fueled by the ever-increasing social awareness of employees and citizens and the scarcity of natural resources, aspects that open great opportunities for the creation of shared value" (Magatti, 2017). The economy we need is a precious building site for making a greater plurality of goods available, raising the living conditions of all, communication of people and expanding the possibilities of expression. (Stiegler, 2010; 2011; 2016; Magatti, Giaccardi 2014, Butera, 2020).

Generativity moves in this direction, in fact it is profiled as a way of individual and social being that seeks to promote, through care, every form of life, developing the ability to give a contribution to the community. The significant contribution of Erikson (1982), mainly focused on the individual characteristics of the generative personality, has allowed us to outline a broader construct related to social generativity, indeed "Marcia (2010) argues that generativity ranges from being individually centered to being socially centered" (Morselli, Passini, 2015, p. 174). Social generativity is a new way of thinking and acting, that talks about the possibility of an action which is creative, responsible, productive, socially oriented and, capable to start a new life: the continuation of society in the future is the basic stage of social generativity (Morselli, Passini 2015). Quoting Morselli and Passini (2015, p.174) "Generativity is closely related to future-related concepts that pinpoint the way people project themselves into the future, such as the future time perspective (Lewin 1951; Zimbardo, Boyd, 1999) and the consideration of future consequences (Strathman et al., 1994)", as well as of social responsibility (Morselli, 2013). Aubin and McAdams (1995) linked generativity to civic sense and to the feelings of attachment to the community (Morselli, Passini 2015). In this sense seems possible to argue that generativity is linked to a definition of community as an inclusive playground for different people and social groups (Morselli, Passini 2015). Generativity is a way of being that is not closed in itself, but that recognizes itself in relationships, open to others and to life. Generativity also requires the creation of conditions for social, human and economic develompement (Monaci, Magatti 2017).

Among these, there is the competence to act and live, derived from the set of cognitive and relational resources that people have, in union with the ability to express themselves (Sen, Naussbaum). In this sense, **Generativity is the development of "capabilities", namely "Capacitation", wich means investing in people to make them capable, competent** (Magatti, Morselli, 2014). This aspect is linked to issues of education and instruction in an increasingly technological and culturally advanced world (Magatti, Morselli, 2014). Only those who are capable, that is to say aware of knowing how to do something, are active in a generative exchange. In this sense, the **rethinking of training and educational paths is a strategic opportunity in the creation of shared value** (Stiegler, 2011; 2019). Equipping individuals and groups with new capacities to act and participate, improving the environment and empowering people and communities/contexts, consciously contributing to cultural change through the testimony of their actions, **generative organizations** invest in new social infrastructures and regenerate common goods, shared meanings, collective identities, individual capacities, sustainable forms of work and production (Magatti, 2017; Magatti Gherardi, 2014a; 2014b; Magatti, Giaccardi, Martinelli, 2019).

According to Magatti and Morselli (2014), generative actions present the following characteristics:

- *Surplus*: a desire for further life, which makes existence continue and ensures progress. Unlike excess, surplus allows life, and therefore society, to flow, renewing the forms of life that are gradually created.
- *Initiative*: it is a concrete act of starting something, of initiating new worlds. It consists in finding new solutions that are better suited to the future that is coming and encouraging its introduction. To generate is not only to be open to the surplus, but to allow it to manifest itself and to be realized by finding the right ground of competence.
- *Valorization*: to generate is to bring value, to help people and things to grow and develop, facilitating their empowerment. In this sense, generativity is understood as the increasing of value in the long term for all the dimensions involved in social and organizational actions. Enhancement is based on an open involvement of all subjects, thinking of future generations.

- *Sustainability*: generativity is being aware that the future can exist only if there are social, economic and environmental conditions for its realization. Sustainability implies a long-term oriented view, which goes beyond the short-term value.

Generativity requires **a return of the value produced, to the territory, to the community** in which one is inserted and to future generations. We do not want to minimize the importance of business and its priorities, the aim is to encourage companies to return and share surpluses with society, allowing a growth of humanity and to ensure a value, including economic, in the long term. From this point of view, generative organizations are institutional forms recognized and legitimized on the basis of their ability to respond creatively, positively and productively to the needs of contemporary human beings (Magatti 2017; Magatti Gherardi 2014; Magatti, Giaccardi, Martinelli 2019; Stiegler 2010).

A Value-Sharing Matrix

But then, what can a company do to be generative, common-oriented in CSV practices? An organization, in a generative key, is a place where **instrumental and human resources are combined to achieve a shared goal of social development and value creation**. Consequently, within the context of this research, we will present **the main areas in which an organization should act to create value, in order to contribute effectively towards the betterment of society** (Staerklé et al., 2015). Authors have been able to identify a **matrix** that is transversal to the different paradigms and the various theoretical contributions taken into consideration, which refer to **multiple levels of citizenship and community life** (Table 1).

Value for Social Citizenship

A first level on which organizations can act is that of social citizenship, promoting a long-term perspective aimed at exchange, generativity and protection of the commons. It consists in commitment to working for the good of larger groups, concern for the community and the next generation, as well as the desire to contribute resources for the long-term good of society (Butera, 2020). It is also about supporting the territory and the community by making individuals capable and competent to act and to live in the world and in relation to others, in an ethical and responsible manner (Coleman, 2001). The main values that support the new citizenship are: continually challenging oneself and going out of one's comfort zone; respecting others and the rules; spending oneself with generosity and commitment; being exemplary and consistent; keeping promises; measuring results objectively; cultivating responsibility and reciprocity; pursuing equity and justice; being transparent and sincere; being open to comparison and exchange; being punctual and respecting time; recognizing and celebrating success (Fregnan, Ivaldi, Scaratti, 2020).

Value for Organizational Citizenship

An organization does not share value just with the outside world and the society. A further action plan concerns the internal dimension, including workers and managers who live the organization on a daily basis (Assiouras, 2019). Creating value for corporate citizenship means promoting initiatives of genuine responsibility towards one's own employees, helping them to recognize themselves as an integral part of the organization (Mirvis, Googins, 2006). It also means supporting workers in their personal and professional growth, making them competent (Pais, Ponzellini, 2020) and stimulating their identification, acknowledgement and gratification. In this sense, the creation of value lies in promoting a sense of engagement and orientation in one's own professional and personal path (Pies, Beckmann, Hielscher, 2010). A real commitment in this direction is able to produce value through a brand improvement.

Value for Employability

A generative conception of value for the company consists of going beyond the maximization of profit, envisages the enhancement of professional skills and capabilities and the evolution of organizational culture (Van den Broeck, 2014). A company creates value through the empowerment of individuals, both its own employees and the members belonging to the community in which it operates (Andrews, Russell, 2012). It is a matter of making individuals capable of working and acting in a long-term perspective, technically competent and able to fit into the various organizational contexts. In this sense, the organization assumes responsibility for supporting the growth of its own resources and of the wider population, developing capabilities at all levels (Baker, Henson, 2010; Wibrow, 2011).

Value for Digital Citizenship

Generativity requires creating the conditions for economic, social and human growth. Among these, there is the "competence to act" (Magatti, 2017), derived from the set of cognitive and relational resources that a person has, in union with his or her ability to express them. (Sen, Naussbaum, 2003). This aspect is linked to issues of learning and schooling in an increasingly technological and culturally advanced world. Only those who are capable, that is, aware of knowing how to do something, are more active in a generative exchange. In this sense, the rethinking of training and educational paths is a strategic opportunity in the creation of shared value (Russo-Spena et al., 2019). The commitment of organizations must be directed towards the development of skills linked to mastery of the technological innovations that characterize the contemporary world (Dovey, Alevizou, Williams, 2016; Matthyssen 2019). This aspect will be further explored in Chapter 2.

Value for Innovation

Value creation is based on long-term orientation, the ability to look ahead and to anticipate change. The capacity to promote innovation is an opportunity for organizations to promote value too. It is about promoting new production processes, new market concepts, new forms of employees' involvement, as well as introducing more advanced technologies (Gummesson et al., 2010). It is about creating new ecosystems of exchange and sharing with the territory and other organizational agents (Matthyssen, 2019). A particular field of

Table 1. Value-sharing matrix

Value for Social Citizenship	**Value for Organizational Citizenship**	**Value for Employability**	**Value for Digital Citizenship**	**Value for Innovation**
Being responsible citizens, oriented towards the future and the development of both the territory and the community.	Being members of the "corporate citizenship", sharing its recognized values and feeling.	Developing skills to work in the world of today and of the future.	Being "digital citizens", able to consciously use technologies and to seize the opportunities offered by the Digital Transformation.	Being able to innovate and to develop continuously, with a view on long-term (sustainable) progress.

innovation is that of didactics. In this text, in fact, we argue that organizations play a central role in promoting a change in learning processes, centered on the creation of skills and the preparation for the world of work of the future (Kim, Mauborgne 1997; 2004).

CONCLUSION

The chapter presented an overview of how the concept of value has evolved. Starting from the theory of shared value of Porter and Kramer and passing through Critical Management Studies. At the end, a critical perspective has been offered, underlining how value sharing goes beyond aspects of profit and competitive edge, placing organizations and HRM functions at the center of the increasing challenges of authentic Corporate Social Responsibility, primarily oriented by an ethical bottom line (Husted, Allen, 2000; Cunliffe, 2021) and the principles of commons (Ostrom, 1990; De Angelis, 2003; De Angelis, Harvie, 2014) and generativity (Magatti, 2017; Stiegler, 2009).

This first overview is useful to provide a theoretical reference framework from which the reflections conducted in this research work take their cue.

REFERENCES

Alvesson, M. (1996). Leadership studies: From procedure and abstraction to reflexivity and situation. *The Leadership Quarterly*, *7*(4), 455–485. doi:10.1016/S1048-9843(96)90002-8

Alvesson, M., & Deetz, S. (2000). Doing critical management research. *Sage (Atlanta, Ga.)*.

Alvesson, M., Hardy, C., & Harley, B. (2008). Reflecting on reflexivity: Reflexive textual practices in organization and management theory. *Journal of Management Studies*, *45*(3), 480–501. doi:10.1111/j.1467-6486.2007.00765.x

Alvesson, M., & Kärreman, D. (2007). Constructing mystery: Empirical matters in theory development. *Academy of Management Review*, *32*(4), 1265–1281. doi:10.5465/amr.2007.26586822

Alvesson, M., & Willmott, H. (1992). On the idea of emancipation in management and organization studies. *Academy of Management Review*, *17*(3), 432–464. doi:10.5465/amr.1992.4281977

Amhøj, C. B. (2019). Sustainability as an ecology of learning, thinking and acting: Using the World Health Organization's six P's as an action-research intervention to create public value with multiple bottom lines. *Local Economy*, *34*(5), 439–455. doi:10.1177/0269094219847124

Andrews, G., & Russell, M. (2012). Employability skills development: strategy, evaluation and impact. *Higher Education, Skills and Work-Based Learning*, *2*(1).

Assiouras, I., Skourtis, G., Giannopoulos, A., Buhalis, D., & Koniordos, M. (2019). Value co-creation and customer citizenship behavior. *Annals of Tourism Research*, *78*, 102742. doi:10.1016/j.annals.2019.102742

Aubin, E. D. S., & McAdams, D. P. (1995). The relations of generative concern and generative action to personality traits, satisfaction/happiness with life, and ego development. *Journal of Adult Development*, *2*(2), 99–112. doi:10.1007/BF02251258

Baiyere, A., Salmela, H., & Tapanainen, T. (2020). Digital Transformation and the new logics of business process management. *European Journal of Information Systems*, *29*(3), 238–259. doi:10.1080/0960085X.2020.1718007

Baker, G., & Henson, D. (2010). Promoting employability skills development in a research-intensive university. *Education+ Training*, *52*(1).

Barney, J. (1991). Firm resources and sustained competitive advantage. *Journal of Management*, *17*(1), 99–120. doi:10.1177/014920639101700108

Barr, R. (2019). Growing up in the digital age: Early learning and family media ecology. *Current Directions in Psychological Science*, *28*(4), 341–346. doi:10.1177/0963721419838245 PMID:31423053

Baxter, R., & Richard, B. (1994). *Baxter: A Holy Commonwealth*. Cambridge University Press. doi:10.1017/CBO9781139170260

Blumler, J. G., & Coleman, S. (2001). *Realising democracy online: A civic commons in cyberspace* (Vol. 2). Citizens Online.

Bollier, D. (2010). *Academia as a Commons*. Commons Magazine.

Bollier, D., & Helfrich, S. (Eds.). (2014). *The wealth of the commons: A world beyond market and state*. Levellers Press.

Bradley, C. L., & Marcia, J. E. (1998). Generativity-stagnation: A five-category model. *Journal of Personality*, *66*(1), 39–64. doi:10.1111/1467-6494.00002

Butera, F. (2020). *Organizzazione e società*. Marsilio.

Centemeri, L. (2018). Commons and the new environmentalism of everyday life. Alternative value practices and multispecies commoning in the permaculture movement. *Rassegna Italiana di Sociologia*, *59*(2), 289–314.

Crane, A., Palazzo, G., Spence, L. J., & Matten, D. (2014). Contesting the value of "creating shared value". *California Management Review*, *56*(2), 130–153. doi:10.1525/cmr.2014.56.2.130

Cunliffe, A. L. (2021). A very short, fairly interesting and reasonably cheap book about management. *Sage (Atlanta, Ga.)*.

De Angelis, M. (2003). Reflections on alternatives, commons and communities. *The Commoner*, *6*(Winter), 1–14.

De Angelis, M. (2006). The World Social Forum: Challenging empires. *Development*, *49*(2), 125–128. doi:10.1057/palgrave.development.1100264

De Angelis, M. (2017). *Omnia sunt communia: On the commons and the transformation to postcapitalism*. Bloomsbury Publishing. doi:10.5040/9781350221611

De Angelis, M., & Harvie, D. (2013). *The commons*. Routledge.

De los Reyes, G. Jr, & Scholz, M. (2019). The limits of the business case for sustainability: Don't count on 'Creating Shared Value'to extinguish corporate destruction. *Journal of Cleaner Production*, *221*, 785–794. doi:10.1016/j.jclepro.2019.02.187

Dembek, K., Singh, P., & Bhakoo, V. (2016). Literature review of shared value: A theoretical concept or a management buzzword? *Journal of Business Ethics*, *137*(2), 231–267. doi:10.100710551-015-2554-z

Dovey, J., Alevizou, G., & Williams, A. (2016). Citizenship, value and digital culture. In I. Hargreaves & J. Hartley (Eds.), The Creative Citizen Unbound, edited by Ian Hargreaves and John Hartley (pp. 75–102). Academic Press.

Erikson, E. H. (1982). *The life cycle completed*. Norton.

Fregnan, E., Ivaldi, S., & Scaratti, G. (2020). HRM 4.0 and New Managerial Competences Profile: The COMAU Case. *Frontiers in Psychology*, *11*, 11. doi:10.3389/fpsyg.2020.578251 PMID:33329228

Galuppo, L., Gorli, M., Scaratti, G., & Kaneklin, C. (2014). Building social sustainability: multi-stakeholder processes and conflict management. *Social Responsibility Journal.*

Gherardi, L., & Magatti, M. (2014). *Una nuova prosperità: Quattro vie per una crescita integrale*. Feltrinelli Editore.

Gummesson, E., & Mele, C. (2010). Marketing as value co-creation through network interaction and resource integration. *Journal of Business Market Management*, *4*(4), 181–198. doi:10.100712087-010-0044-2

Hess, M. C. T., & Benlian, A. (2015). Digital Transformation strategies. *Business & Information Systems Engineering*, *57*(5), 339–343. doi:10.100712599-015-0401-5

Hillman, A. J., & Keim, G. D. (2001). Shareholder value, stakeholder management, and social issues: What's the bottom line? *Strategic Management Journal*, *22*(2), 125–139. doi:10.1002/1097-0266(200101)22:2<125::AID-SMJ150>3.0.CO;2-H

Husted, B. W., & Allen, D. B. (2000). Is it ethical to use ethics as strategy? In *Business challenging business ethics: New instruments for coping with diversity in international business* (pp. 21–31). Springer. doi:10.1007/978-94-011-4311-0_4

Husted, B. W., & Allen, D. B. (2011). *Corporate social strategy*. Academic Press.

Ketchen, D. J. Jr, & Hult, G. T. M. (2007a). Bridging organization theory and supply chain management: The case of best value supply chains. *Journal of Operations Management*, *25*(2), 573–580. doi:10.1016/j.jom.2006.05.010

Ketchen, D. J. Jr, Hult, G. T. M., & Slater, S. F. (2007b). Toward greater understanding of market orientation and the resource-based view. *Strategic Management Journal*, *28*(9), 961–964. doi:10.1002mj.620

Kim, W. C., & Mauborgne, R. (1997). Value Innovation: The Strategic Logic of High Growth. *Harvard Business Review*, *75*(1), 102–112. PMID:10174449

Kim, W. C., & Mauborgne, R. (2004). Value innovation. *Harvard Business Review*, *82*(7/8), 172–180. PMID:10174449

Lewin, K. (1951). *Field theory in social science: selected theoretical papers* (D. Cartwright, Ed.).

Magatti, M. (2017). *Cambio di paradigma. Uscire dalla crisi pensando il futuro*. Feltrinelli.

Magatti, M., & Giaccardi, C. (2014). *Generativi di tutto il mondo unitevi! Manifesto per la società dei liberi*. Feltrinelli.

Magatti, M., Giaccardi, C., & Martinelli, M. (2019). Social generativity: A relational paradigm for social change. In *Große Transformation? Zur Zukunft moderner Gesellschaften* (pp. 469–486). Springer VS. doi:10.1007/978-3-658-25947-1_26

Matinheikki, J., Rajala, R., & Peltokorpi, A. (2017). From the profit of one toward benefitting many–Crafting a vision of shared value creation. *Journal of Cleaner Production, 162*, S83-S93.

Mirvis, P., & Googins, B. (2006). Stages of corporate citizenship. *California Management Review, 48*(2), 104–126. doi:10.2307/41166340

Monaci, M., & Magatti, M. (2017). Generative dynamics: What sustains the creation of shared business value. In *Technology, Society and Sustainability* (pp. 373–395). Springer. doi:10.1007/978-3-319-47164-8_27

Morselli, C. (Ed.). (2013). *Crime and networks*. Routledge. doi:10.4324/9781315885018

Nussbaum, M. (2003). Capabilities as fundamental entitlements: Sen and social justice. *Feminist Economics, 9*(2-3), 33–59. doi:10.1080/1354570022000077926

Opotow, S. (1990). Deterring moral exclusion. *The Journal of Social Issues, 46*(1), 173–182. doi:10.1111/j.1540-4560.1990.tb00280.x

Opotow, S. (1990a). *Moral exclusion and injustice: An introduction*. Academic Press.

Opotow, S. (2008). "Not so much as place to lay our head...": Moral Inclusion and Exclusion in the American Civil War Reconstruction. *Social Justice Research, 21*(1), 26–49. doi:10.100711211-007-0061-9

Ostrom, E. (1990). *Governing the commons: The evolution of institutions for collective action*. Cambridge university press. doi:10.1017/CBO9780511807763

Ostrom, E., Gardner, R., Walker, J., Walker, J. M., & Walker, J. (1994). *Rules, games, and common-pool resources*. University of Michigan Press. doi:10.3998/mpub.9739

Pais, I., & Provasi, G. (2015). Sharing economy: A step towards the re-embeddedness of the economy? *Stato e mercato, 35*(3), 347-378.

Passini, S. (2010). Moral reasoning in a multicultural society: Moral inclusion and moral exclusion. *Journal for the Theory of Social Behaviour*, *40*(4), 435–451. doi:10.1111/j.1468-5914.2010.00440.x

Passini, S., & Morselli, D. (2010). The obedience–disobedience dynamic and the role of responsibility. *Journal of Community & Applied Social Psychology*, *20*(1), 1–14. doi:10.1002/casp.1000

Pies, I., Beckmann, M., & Hielscher, S. (2010). Value creation, management competencies, and global corporate citizenship: An ordonomic approach to business ethics in the age of globalization. *Journal of Business Ethics*, *94*(2), 265–278. doi:10.100710551-009-0263-1

Porter, M., & Kramer, M. R. (2011). Creare valore condiviso. *Harvard Business Review Italia*, *1*(2), 68–84.

Porter, M. E., & Kramer, M. R. (2019). Creating shared value. In *Managing sustainable business* (pp. 323–346). Springer. doi:10.1007/978-94-024-1144-7_16

Prasad, A., & Holzinger, I. (2013). Seeing through smoke and mirrors: A critical analysis of marketing CSR. *Journal of Business Research*, *66*(10), 1915–1921. doi:10.1016/j.jbusres.2013.02.013

Rocha, C. S., Antunes, P., & Partidário, P. (2019). Design for sustainability models: A multiperspective review. *Journal of Cleaner Production*, *234*, 1428–1445. doi:10.1016/j.jclepro.2019.06.108

Russo-Spena, T., Mele, C., & Marzullo, M. (2019). Practising value innovation through artificial intelligence: The IBM watson case. *Journal of Creating Value*, *5*(1), 11–24. doi:10.1177/2394964318805839

Staerkle, C., Falomir-Pichastor, J. M., Pereira, A., Berent, J., & Butera, F. (2015). Global value perceptions: The legitimising functions of western representations of democracy. *European Journal of Social Psychology*, *45*(7), 896–906. doi:10.1002/ejsp.2159

Stiegler, B. (2009). The carnival of the new screen: From hegemony to isonomy. *The YouTube Reader*, 40-59.

Stiegler, B. (2010). For a new critique of political economy. *Polity*.

Stiegler, B. (2010a). *Taking care of youth and the generations*. Stanford University Press.

Stiegler, B. (2011). *Decadence of industrial democracies* (Vol. 1). Polity.

Stiegler, B. (2016). *Automatic society, volume 1: The future of work*. Academic Press.

Stout, S. S., Simpson, L. A., & Singh, P. (2019). Trust between health care and community organizations. *Journal of the American Medical Association*, *322*(2), 109–110. doi:10.1001/jama.2019.1211 PMID:31180461

Strathman, A., Gleicher, F., Boninger, D. S., & Edwards, C. S. (1994). The consideration of future consequences: Weighing immediate and distant outcomes of behavior. *Journal of Personality and Social Psychology*, *66*(4), 742–752. doi:10.1037/0022-3514.66.4.742

Strathman, A., Gleicher, F., Boninger, D. S., & Edwards, C. S. (1994). The consideration of future consequences: Weighing immediate and distant outcomes of behavior. *Journal of Personality and Social Psychology*, *66*(4), 742–752. doi:10.1037/0022-3514.66.4.742

Van Hiele, A., Mervielde, I., & De Fruyt, F. (2006). Stagnation and generativity: Structure, validity, and differential relationships with adaptive and maladaptive personality. *Journal of Personality*, *74*(2), 543–574. doi:10.1111/j.1467-6494.2006.00384.x PMID:16529586

Vial, G. (2019). Understanding digital transformation: A review and a research agenda. *The Journal of Strategic Information Systems*, *28*(2), 118–144. doi:10.1016/j.jsis.2019.01.003

Westerman, G., Bonnet, D., & McAfee, A. (2014). *Leading digital: Turning technology into business transformation*. Harvard Business Press.

Wibrow, B. (2011). Employability Skills. At a Glance. National Centre for Vocational Education Research Ltd.

Zimbardo, P. G., & Boyd, J. N. (1999). Putting time in perspective: A valid, reliable individual-differences metric. *Journal of Personality and Social Psychology*, *77*(6), 1271–1288. doi:10.1037/0022-3514.77.6.1271

ENDNOTE

1 The authors of this section are Leonardo Ciocca, Università Cattolica del Sacro Cuore di Milano; Ezio Fregnan, Comau Academy & Education Business Director; Francesca Grignolio, Consultant for Comau Consultancy & Education Sector.

Chapter 2
Digital Transformation Scenario, Company Evolution, and Request for New Skills

ABSTRACT

We live in a world characterized by technological development that is proceeding at exponential speed and that has brought about substantial changes at a social and organizational level. The effects of the Fourth Industrial Revolution and the digital transformation require individuals and organizations to adapt, developing new skills and a new work culture capable of accompanying the changes taking place and empowering the workforce in this new scenario. The digital transformation does not only imply physical or structural changes related to new technological devices; it also requires the development of new ways of working and of relating to new technologies and to other individuals.

INTRODUCTION

This chapter presents the historical and social scenario in which the research of this book takes place, a context characterized by radical technological innovations (Hess, Benlian, 2015) that have triggered profound changes in the way we work, relate to others and live in the world and in the communities we belong to (Vial, 2019).

DOI: 10.4018/978-1-7998-8653-2.ch002

The starting hypothesis is **that in this condition there are new spaces to create social value, impacting on both organizational dimensions and everyday life issues** (Westerman, Bonnet, McAfee, 2014). Innovation entails great opportunities, but at the same time significant risks at individual, corporate and community levels (Verhoef, Broekhuizen, Bart, Bhattacharya, Dong, Fabian, Haenlein, 2021). Popper (1994) defines knowledge as a way of solving a problem by introducing innovative solutions capable to fulfill a specific need. However, once the starting problem has been solved, the changes introduced lead to new critical issues that did not exist before and need to be solved through the development of new knowledge. The authors uses an example to clarify this concept: in order to solve the problem of long distance, the human being has invented the car, so that he could move faster. On the one hand, the introduction of the car has effectively solved the problem of distance, but on the other hand, it has led to the emergence of new criticalities (previously non-existent and unpredictable ones) related to this innovation, such as traffic, pollution and parking (Popper, 1994). Parallel to this concept, the radical technological innovations can be understood as a form of progress (based on a knowledge development) designed to solve problems of productive efficiency and effectiveness (Verhoef et al. 2021), but at the same time it generates implications that need to be managed with new knowledge and through the development of specific skills (Warner, Wager, 2019). The management of these elements must pass through the **development of technical, digital, social and new labor skills**, and **facilitating the strengthening of these skills is a real form of value creation** (Van Laar, Van Deursen, Van Dijk, De Haan, 2017).

By means of a desk-based research (Guerin, Janta, Van Gorp, 2018), in the following pages authors will discuss the concepts of Fourth Industrial Revolution and Digital Transformation, outlining their main characteristics, organizational and societal implications, opportunities and risks implied and, finally, the skills that will be required for working in the future. If, as Popper (1994) argues, knowledge production is the answer to a problem, what is the starting point of this desk-based research? Different methodological steps led to the creation of the chapter.

The world in which we live is characterized by a condition of disruptive change and uncertainty (Ismail, Khater, Zaki, 2017), an aspect made even more evident by the radical technological innovations we are witnessing. The challenge is to promote the creation of value within a scenario of technological innovation, impacting both the society in which we live and the world of organizations (Reddy, Reinartz, 2017). The chapter aims to **describe the**

context of digital and technological innovation and its implications at social and organizational level, to **identify then the skills required to cope with such a transformative scenario** (Hoberg, Krcmar, Welz, 2017). Since this is still an evolving phenomenon, authors have decided to describe it from a theoretical point of view, in order to make a positioning choice and to establish a knowledge starting point. In adherence with these goals and their nature, authors identified **desk-based research as the most functional research method** (Guerin, Janta, Van Gorp, 2018); This secondary research approach does not involve the collection of empirical material from the field, but rather the study of existing literature, with the aim of identifying the **theoretical anchors to understand and describe our phenomenon of interest**, namely the **Digital Transformation scenario**, the **company evolution** and the **skills that will be necessary** to adequately manage the future challenges (Guerin, Janta, Van Gorp, 2018).

Initially, data collection included an in-depth literature review on the concepts of "Digital Transformation Scenario", "Company Evolution" and "Request for New Skills". This step allowed to identify **recurring elements for data processing**, which was accomplished by defining a set of categories and keywords of reference. The research phenomenon has been explored through the **constructs of Fourth Industrial Revolution and Digital Transformation, implications on organizations and society and demand for new skills**. These elements will be presented in the next pages.

DATA COLLECTION AND PROCESSING

Fourth Industrial Revolution

The Fourth Industrial Revolution refers to the increasing interpenetration of the physical, digital, and biological worlds (Fregnan et al., 2020). The disruptive innovation driven by the Fourth Industrial Revolution concerns the thinning of boundaries between the physical, digital and biological spheres (Shahroom, Hussin, 2018). Robots and Social Media (SM) are just some few components of Fourth Industrial Revolution technology source (Yusuf, Walters, Sailin, 2020), portrayed by 3 primary dimensions: the inexorably increasing utilization of Internet, the improvement and application of Artificial Intelligence, the dissemination of automatic learning (Schwab, 2016; Fregnan et al., 2020). It is a sum of development in artificial intelligence (AI), robotics, Internet of

Things (IoT), 3D printing, genetic engineering, quantum computers and other technologies. It is the collective force behind many products and services that are rapidly becoming essential to modern life (Lee, Lee, Chou, 2017).

The technological development does not imply just significant structural innovations in industrial realities, but also a transformation of markets and society as a whole (Morrar, H. Arman, S. Mousa, 2017). Just think about how the introduction of the steam engine, a technological artifact whose introduction symbolizes the beginning of the First Industrial Revolution, led to the translation of the logic of artisan production into the principles of mass production that laid the foundations for contemporary society, evolving into the more complex forms of Taylorism, Fordism and capitalism (Pereira, Romero, 2017). In addition to these aspects, we can see clearly how the birth of the first factories triggered demographic and social movements, such as the great phenomena of urbanization and migration from the countryside that first involved the English cities and then spread to the rest of the industrialized countries, until the birth of a new social class, the working one, putting the basis for new problems, still relevant today, as workers' rights, social justice, welfare, etc. (Lu, 2017). Two and a half centuries have passed since the introduction of that technology, and two industrial revolutions happened after the first one; the world has experienced exponential growth in terms of culture, globalization and well-being, at the same time the expansion of the world's boundaries has meant greater complexity and a sense of uncertainty never before experienced in this era, all accompanied by widespread awareness among individuals and an increasing demand for corporate responsibility (Li, Hou, Wu, 2017).

These elements of social character, related to the possibility of a technological improvement unpredictable until a few years ago, necessarily require **considering the impacts of the Fourth Industrial Revolution and the Digital Transformation on society** (Li et al., 2017), as it is believed that it may represent both an opportunity for sustainable progress at all levels and a possible risk to collective and individual well-being (Vey, Fandel-Meyer, Zipp, Schneider, 2017). Nowadays, we are experiencing a period of intense changes that are difficult to ascribe to a single field or sector. Driven by the Fourth Industrial Revolution, we are entering a new period of development in which digital, physical and biological technology have reached an unprecedented development both in their individual fields and in mutual interaction (Guoping, Yun, Aizhi, 2017). The disruptive innovation brought by this latest revolution is, in fact, a way to thin the boundary between the physical, digital and biological spheres with the potential to radically change how we live: from the way we work to the way we perceive the interaction with others (Schwab, 2016). The increasing usage of digital technologies brings forward a huge amount of major changes in both business and society. Often, in literature, the term **Fourth Industrial Revolution** is used as a synonym of **Industry 4.0** (Hecklau et al., 2016), a concept introduced in Germany during the 2011 Hannover Fair and concerning the country's high-tech strategy for 2020. However, considering the broader scope of the Fourth Industrial Revolution that does not only affect the industrial domain but the entire socioeconomic system, it should be important to consider them as **two certainly dependent but separate concepts** (Philbeck, Davis, 2018).

The Fourth Industrial Revolution: An Overview

Philbeck and Davis (2018) describe the Fourth Industrial Revolution as a set of major changes in the way in which economic, political and social value is created, exchanged and distributed, closely related to the idea behind the Revolution itself, namely that the digital, physical and biological worlds combine and reinforce each other. In fact, the core of this revolution is often identified in the Cyber-Physical Systems (CPS), "technologies that create integration between virtuality and reality by creating a network in which intelligent objects communicate with each other" (Li, Hou, Wu, 2017). At this time, it is still difficult to fully define what this Revolution is and which effects implies. We certainly know, however, that it is affecting every field of human life touching all aspects of society and economy (Li, Hou, Wu, 2017).

In general, industrial revolutions are moments of technological change linked to broader social transformations (Philbeck, Davis, 2018). Each industrial revolution has brought economic growth, has increased productivity and prosperity, but, at the same time, challenges too (Piccini et al., 2015). The current Revolution, likewise, is expected to bring **great benefits but also intense moral and ethical challenges** (Svahn et al., 2017) related, for example, to a possible increase in inequalities associated with new levels of skills required for work, higher levels of automation and human-robot interaction, as well as concerning privacy. All this calls for a **holistic approach** to the treatment of the Fourth Industrial Revolution, managing it by considering the **three fundamental dimensions of sustainable development**: the economic dimension, the social dimension and the environmental dimension (Morrar, Arman, Mousa, 2017; Amhoj 2019). Schwab (2016) also points out that what distinguishes this Revolution from the previous ones, and in particular from the third one with which it shares the technological base, is: the **speed of its evolution**, exponential rather than linear, the **scope of the changes** introduced and the **unprecedented impact** on social, organizational and economic systems.

The Fourth Industrial Revolution: Key Technological Drivers

Li Guoping, Hou Yun, Wu Aizhi (2017) identify **three key technological drivers** of the Fourth Industrial Revolution:

1. *Digital driver.* This is the center of Fourth Industrial Revolution and Industry 4.0. The digital driver is the one that empowers, and makes conceivable, practically the entirety of the advances presented by this Revolution. The most pertinent innovations in this sense are the Internet of Things (IoT), Artificial Intelligence (AI), Big Data and Cloud Computing and the Digital Platform. IoT is based on the thought of causing unavoidable presence of things around us able to speak with one another to accomplish shared objectives (Li et al., 2017). This innovation, empowering Smart Products, allows them to gather data, collaborate with their environmental factors, yet in addition to be associated with different items with which to trade information and data using web (Pereira, Romero, 2017). Artificial Intelligence is an innovation whose advancement started during the 1990s, it permits computers to be outfitted with programmed learning measures, recreating conduct in a way that allows individuals to learn from the available data and make reliable and repeatable decisions independently. Big Data and Cloud Computing are identified with the management and the processing of the vast amount of available data. Digital Platforms link individuals to each other, to assets and information, creating another method of conveying and devouring administrations, taking into account more noteworthy association in the organic market (Li et al., 2017).
2. *Physical Driver*. The authors recognize two fundamental components identified with this driver: Autonomous vehicles and 3D printing. The first one brings various benefits identified with crash decrease, discharge decrease with fuel investment funds, and could likewise permit versatility for aged and handicapped individuals who might be blocked from driving. 3D printing or added substance fabricating, could again empower cost decrease and reduce lead times (Li et al., 2017).

3. *Biological Driver.* According to the authors, this is the latest driver of the Fourth Industrial Revolution and is related to neurotechnologies and genetic technology. In particular, advances in neurotechnologies are leading to improved performance of robots, that will have an increasingly pervasive role in society and in the interaction with humans (Li et al., 2017).

The Fourth Industrial Revolution introduces radical changes at technological level; consequently, their application in business broadens the horizon to the **construct of Digital Transformation**.

Digital Transformation

Digital Transformation is the **fast development of technology to perform business exercises, upgrading the performance, events, practices, and models in reacting to the rising effects and challenges of technological innovation (Gebayew et al., 2018)**. It likewise allows companies to improve operational efficiencies, customer experience, and the combination of digital and physical business (Van Valdhoven, Vanthienen, 2021). Then again, Digital Transformation is the disruptive and quick change of business exercises, processes, capabilities and models to fully leverage the challenges and opportunities of advanced innovations and their effect across society with a strategic approach (Vial, 2019; Wade, 2015). Furthermore, this is the utilization of new digital tools (social media, mobile, analytic software) to trigger significant business upgrades (Westerman, Bonnet, 2015), for example, improving the customer experience (Von leipzig et al., 2017) or implementing new action plans (Matt et al., 2015).

Van Veldhoven and Vanthienen (2020) highlight that the idea of Digital Transformation was initially introduced in 2000 (Patel and McCarthy, 2000) but it was only after 2014 that the term quickly gained notoriety by both practitioners and scholars (Reis, Amorim, Melão, and Matos, 2018). The authors (Veldhoven and Vanthienen, 2020) clarify that various and dissimilar definitions have been proposed for Digital Transformation, starting from a connection that hopes to improve a substance by setting off critical changes, to describing its properties through blends of data, processing, correspondence and availability advancements (Vial, 2019), until the improvement that the technological innovation causes or its effects in all dimensions of human life (Stolterman, Fors, 2004).

It is worth specifying that **Digital Transformation is not a completely new idea**; its foundations can be found in digitization, turning simple information and processes into digital innovation (Henriette, Feki, Boughzala, 2015; Veldhoven and Vanthienen, 2021), and in digitalization, the sociotechnical interaction in which advances are embraced at a large scale (Legner et al., 2017; Veldhoven and Vanthienen, 2021). There is some semantic confusion encompassing these definitions, in any case Digital Transformation is outlined as a more extensive change, if compared to digitalization (Veldhoven and Vanthienen, 2021). By examining the Digital Transformation definitions, it is possible to split them into the most common key dimensions: **society change, utilization of digital technologies, customer experience, change measure, value creation, new action plans, hierarchical change, digital, advanced economy, interior activities, organizational transformation, products and services** (Vial 2019; Fitzgerald, Kruschwitz, Bonnet, Welch, 2014). A Digital Trasformation strategy aims to **create the capabilities** within the organization, in order to take full advantage of the opportunities of new technologies and their impacts in the best, fastest, and most innovative way (Gebayem et al. 2018). (https://www.i-scoop.eu/digital-transformation/)

The Three Axes of Digital Trasformation: Changes in Digital Technologies, Business, and Society

Under the Digital Transformation dimension, Veldhoven and Vanthienen (2021) include the **major technology improvements** that have guided the change over long time, which can be summed up in five waves dependent on the historical evolution (Veldhoven and Vanthienen, 2021):

1. The first wave began with the beginning of IT innovation from automation of simple activities through the use of PC innovation (Legner et al., 2017; Venkatraman, 1994).
2. Afterward, web or ICT answers for quicker data sharing and network became predominant (Heavin and Power, 2018; Legner et al., 2017).
3. With the massive rise in popularity of digital technologies and the web, more reconciliation arrangements showed up, like applications to interface a lot of information, clients, and PCs in smart software and technology solutions (Heavin; Power, 2018).
4. The fourth and current wave, as highlighted by many models, involves the massive ascent of online media, mobiles, analytics, cloud computing and the Internet of Things (Chanias, 2017; Legner et al., 2017; Vial, 2019)

5. In the next wave Veldhoven and Vanthienen (2021) include arising advances new technologies whose business potential is not fully exploited yet, such as blockchain, mechanical technology, 3D printing, AI, and cognitive computing (Schwab, 2015; Verhoef et al., 2019).

These five categories catch the main advanced innovations that assume an indispensable role in DT.

According to Veldhoven and Vanthienen (2021), the business dimension addresses **the most significant changes occurring over organizations because of the expanded impact of digital technologies and the digitalized society**. The authors base themselves on the apparent development and shared characteristics emerging from the desk-based research under five levels (Veldhoven and Vanthienen, 2021):

1. Organizations began to digitize or automate inner business measures and exercises (Venkatraman, 1994).
2. After that, the network between various business measures, providers, frameworks or customers changed from manual and physical to digital (Morgan and Page, 2008).
3. The next change occurred in the contribution of business. Items and administrations were digitalized and coordinated to follow the rapidly changing requirements and requests (Hartl and Hess, 2017; Henriette et al., 2015; von Leipzig et al., 2017).
4. The expanded speed of DT urged organizations to begin changing their organizational hierarchy aiming to be more flexible, agile and ready to quickly react to changes in the market (Kaufman and Horton, 2014; Vial, 2019).
5. Finally, as the principles of the business game got changed, organizations progressively improved their business plan (Berghaus, Back, 2016; von Leipzig et al., 2017). Changes in strategy concern eco-systems (Subramaniam, Iyer, Venkatraman, 2019), creation of new strategical partnerships (He, Meadows, Angwin, Gomes, Child, 2020) and moving from production to service development (Frank, Mendes, Ayala, Ghezzi, 2019).

Digital Technologies Transformation. Many digital technologies that are driving Digital Transformation can be accredited, of which five significant waves in the previous 50 years are referenced sequentially (Carter, 2018; Porter, Heppelmann, 2014; Singh, Hess, 2017). Even if the last wave, AI, is still in its beginning phases, it is included because many businesses see it as a potential technology for Digital Transformation in the near future (Carter, 2018; HM Government, 2017):

1. IT: automation of isolated activities and problem calculations
2. Internet: inexpensive, ubiquitous connectivity and sharing of information
3. Integration: more affordable, widespread computing power and possibilities
4. SMACIT: social media, mobile, analytics, cloud computing, and Internet of Things
5. Advanced AI: tasks that normally require human intelligence

Business Transformation can be perceived as the progresses in organizations to achieve relevant performance improvement and value generation (McKeown, Philip, 2003). A developing number of parts of the business is changing, which can be sum up in five classifications motivated by the related work (Kane et al., 2017; Venkatraman, 1994; Wade, 2015). The classifications are ranked chronologically, as these dimensions frequently change over time (Morgan and Page, 2008; Venkatraman, 1994).

1. Internals: changes in work, innovation and interior tasks.
2. Connectivity: changes in the association between organizations, processes and entities
3. Offerings: new products and services
4. Structure: changes in the structure, jobs, activities, culture and hierarchy organization
5. Business model: changes in business plan, strategies, goals and management

The business changes are brought by the expanded exploitation and utilization of digital technologies (Venkatraman, 1994; Wade, 2015). The simple use of new technologies for the same way of working in these aspects is not enough; organizations should change their methodologies in these perspectives (Earley, 2014). In an ideal world, the organizations should mean to drive the digital technologies to change all perspectives, and ultimately the business plan (Kane et al., 2017). This issue implies that every part of the business should be improved given the current digital technologies. Note that these viewpoints do not always change in a fixed order. For instance, an organization can be digitalizing its business plan while maintaining its old design. Moreover, the above-mentioned change interaction can be repeated for each new technological innovation wave. Hence, it is a process in which various business aspects can change subsequently and simultaneously with different digital technologies.

Society Transformation can be perceived as the progresses occurring in individuals, customers and society over time (Veldhoven and Vanthienen, 2021). The society axis is centered around the human side of progress. The authors (Veldhoven and Vanthienen, 2021) list the significant implications on people, customers, communities and governments because of the expanded impact of digital technologies and digitalized organizations. In the first place, the daily tasks people perform in their private lives and at their jobs, in line with the business transformation, have considerably changed, if compared to the pre-digital era (Reddy, Reinartz, 2017; Veldhoven and Vanthienen, 2021), in addition, communication has essentially changed. "All of this is changing the possibilities of how to structure our society (Reddy & Reinartz, 2017)" (Veldhoven and Vanthienen, 2021, p.8). The profound effects of DT can also influence people's values, beliefs, privacy ideologies and expectations, (Veldhoven and Vanthienen, 2021; Reddy & Reinartz, 2017; Schwab, 2015). With regards to Digital Transformation, it is about the progresses occurring because of the expanded utilization of digital technologies. People are becoming increasingly digitalized because they adopt digital technologies in every aspect of their lives. This affects their acceptance of digital trends, their identity, their notion of privacy, their work, the way they communicate and the way they live (Hanelt et al., 2015; Schwab, 2015). As customers, their consumption patterns, notion of ownership, demands and product knowledge have intensively changed (Berman, 2012; Schuchmann & Seufert, 2015; Schwab, 2015). As an aggregated effect, this also happens with the entire society (Ebert & Duarte, 2018). These changes can be briefly summarized in five categories. Similarly, to the business changes above mentioned, we argue that these changes usually happen in this order, but they do not have to. Likewise, these changes can happen subsequently and simultaneously with different technologies:

1. Tasks: changes in disengaged errands and exercises
2. Connectivity: changes in correspondence, informedness and network
3. Usage: changes in items and administrations used in everyday life
4. Structure: changes in balance between serious and fun activities, propensities and schedules
5. People: changes in individuals' qualities, idea of possession, character and way of living

Main Implications: Society and Company Evolution

As we have already highlighted above, the scope of the implications related to the Fourth Industrial Revolution goes beyond the boundary of the business sphere alone, also impacting the social and political dimensions (Schwab, 2016). Henriette et al. (2015) showed a systematic literature review on vulnerabilities and opportunities of Digital Transformation, such as on digital capabilities (Warner, Wäger 2019; Karimi, Walter, 2015), business models, customer experience and operational processes (Vial, 2019). We can state that the Fourth Industrial Revolution and the Digital Transformation pose several opportunities and challenges to organizations and societies. We are still at the beginning of this profound change able to invest every aspect of human life, and to understand it and deal with it in the best way, maximizing its benefits and minimizing its costs, it is necessary to adopt a holistic approach that considers the impacts of this phenomenon under the three dimensions of sustainability: the economic, social and environmental ones (Amoji, 2019; Morrar, Arman, Mousa, 2017). As Schwab (2016) argues, neither technology nor the disruptions it has caused are factors that human endure without the possibility of response. **These profound changes need the collaboration of society, organizations and businesses to be directed towards common goals and values**.

Over the next few pages, authors will discuss the **main impact levels** of the radical change we are witnessing, specifically delving into the organizational and social levels.

The Inequality Issue

Schwab (2016) identifies inequality as the most important social issue related to the impact of the Fourth Industrial Revolution and Digital Transformation. In conjunction with the new industrial paradigm, it is in fact causing a trend of substitution of capital for labor to emerge, making the biggest beneficiaries of this process the providers of intellectual and physical capital innovators, investors and shareholders (Guoping, Yun, Aizhi, 2017). This trend would lead to a wealth gap between those who depend on labor and those who depend on capital (Schwab, 2016). The Fourth Industrial Revolution is having a powerful impact on the labor market (Li et al., 2017; Vial, 2019), leading to a strong reduction in the demand for standardized mechanical jobs that will be increasingly automated, as well as a strong increase in demand for technological and digital skills that the labor market is still lacking (Pinto et al., 2021). Therefore, the Fourth Industrial Revolution and the new business paradigm (Li et al., 2017; Vial, 2019) determine a modification of the occupational profiles, of the management and organization of the job, requiring new competences and abilities, with particular reference to the human-robot interface. Schwab (2016) clarify that, in order to avoid the exacerbation of inequalities, the so-called technological unemployment and the emergence of new social tensions, it is necessary to provide profound changes in the current educational framework and to train the current and future workforce to the new market needs, related to the emerging positions that will be created by the Fourth Industrial Revolution (Pereira, Romero, 2017).

Another relevant issue related to the Fourth Industrial Revolution and Industry 4.0 concerns privacy (Schwab, 2016). The new industrial paradigm is based on connectivity, and data transfer and analysis. It is from the analysis of consumer data that it is possible for organizations to meet their needs. Consequently, one wonders about the effects of the loss of control of our data imposed by the new paradigm of connectivity (Morrar, Arman, Mousa, 2017).

Other ethical and moral issues emerging from the Fourth Industrial Revolution concern the increasingly pervasive presence of robots in our lives. The important developments related to neurotechnologies applied to robotics and related to Artificial Intelligence will lead us to new ethical and moral reflections (Schwab, 2016).

Digital Transformation, Social Wellbeing and Healthcare

Several articles also reflect on the impacts of Digital Transformation at higher levels, including the organizational and social dimensions (Vial, 2019). Quoting Vial (2019, p. 1) "Researchers have argued that digital technologies afford a tremendous potential for the improvement of both wellbeing and quality of life of individuals (Agarwal et al. 2010; Pramanik et al., 2016)". One example is healthcare (Agarwal et al., 2010), where different types of technologies are perceived as valuable contributions to a sector that has traditionally been a laggard in the adoption of technology (Vial 2019). Srivastava and Shainesh (2015) studied found that digital technologies enable healthcare organizations to increase access to care, while simultaneously reducing costs for both the organization and patients (Vial, 2019).

Company Evolution and Industry 4.0

The innovations achieved thanks to the Fourth Industrial Revolution and the Digital Transformation have fundamentally modified value chains across all enterprises driving associations to essentially reconsider their business plan, processes and products (Schwab, 2016). The Fourth Industrial Revolution has prompted the rise of another modern worldview: Industry 4.0.

The term Industry 4.0 was first articulated at the Hannover Fair in Germany in 2011, regarding the use of advanced innovations to assembling (Philbeck, Davis, 2011). From that point forward, this idea has got huge consideration. Pereira and Romero (2017) define Industry 4.0 as an umbrella term illustrative of another industrial paradigm that joins the most recent modern improvements identified with Cyber-Physical Systems, Internet of Things, Internet of Service, Augmented Reality and Cloud Manufacturing empowering a more brilliant, more adaptable assembling climate described in terms of higher profitability and proficiency. While there is not yet widespread consensus on all aspects surrounding Industry 4.0, there is agreement on some of its essential features (Pereira, Romero, 2017):

- The term "Smart" assumes a critical role in characterizing the significance of Industry 4.0. The fate of assembling goes toward Smart Factories and Smart Products. The former are manufacturing solutions that provide real-time communication between all manufacturing resources, while the latter are products with a high degree of autonomy that can interact with the physical environment and other objects by sharing information and data. (Pereira, Romero, 2017). Such a smart environment enables important improvements in efficiency, dynamism and flexibility (Lu, 2017);
- The definition of innovative business models. These, along with a more joint effort arranged, empowered by the new advances presented by the Fourth Industrial Revolution and the DT, allow a greater fitting to new organizational and market needs (Pereira, Romero, 2017).
- Another implication on the customer side. As Schwab (2016) states, with the Fourth Industrial Revolution, significant changes have additionally happened on the demand side: customers are requesting a more prominent level of commitment, engagement and communication. The new technologies presented by the Fourth Industrial Revolution, by making the production process more dynamic and flexible, promote

a faster and less expensive response to different customer needs and this could lead to a shift from mass production to mass customization (Rüßmann, Lorenz, Gerbert, Ealdner, Justus, Engel, Harnisch, 2015).

Accordingly, it arises that there are two principal credits to the Industry 4.0: integration and interoperability (Lu, 2017). When we talk about integration in Industry 4.0, we refer to horizontal integration that allows the integration of different IT systems, processes, resources and information within the organization or different organizations, as well as vertical integration that provides, instead, the integration of the same elements but between different organizational hierarchical levels and departments allowing an agile circulation of data along all organizational levels, with the opportunity for a faster adaptation to market changes (Pereira, Romero, 2017). Interoperability, on the other hand, refers to the ability of systems to exchange data and share information and knowledge, an attribute made possible by the use of CPS technologies (Lu, 2017).

This new industrial paradigm, in adherence with the digitization of work processes, collaborative human-robot connection and the use of shared information supported by big data and cloud computing, requires organizations to make significant efforts in terms of both investment and organizational change to address this new competitive benchmark brought by the Fourth Industrial Revolution, that promises significant improvements in productivity, efficiency, flexibility, and speed of reaction to market changes (Bughin, Deakin, O'Beirne, 2019). Associations should confront significant social changes identified with the Digital Transformation and update the abilities of their HR to accomplish the objectives proposed by Industry 4.0 (Vey, Fandel-Meyer, Zipp, Schneider, 2017). In any case, what are the ramifications of the computerized change and the fourth modern transformation advanced by associations?

HRM and Organizational Culture

Digital Transformation in organizations does not only involve the introduction of new technologies and working methodologies, it also introduces challenges that were not considered before and the development of new working habits (Zinder, Yunatova, 2016). Digital Transformation must be accompanied by a cultural change. In this sense it is crucial the role of top management, responsible for defining the kind of culture they want in their organization:

a culture that embraces change and innovation, allows failure, encourages curiosity and keeps the company's mission and core business adherent to the Digital Transformation transformation (Veldhoven, Vanthienen, 2021). Cultural change involves opportunities, risks, and resistance; it is critical that the organization, not only embraces Digital Transformation, but also becomes able to recognize and manage the various challenges (Zinder, Yunatova, 2016). Schein's (2000) model tends to consider organizational culture as a multilevel construct, consisting of artifacts, values and basic assumptions. Artifacts are visible organizational structures and processes; they are what the members of a group, organization, or community see, hear and feel on a daily basis during their activities. Values and basic assumptions, on the other hand, concern beliefs, representations, unwritten rules, and taken-for-granted convictions that determine practices, behaviors, thoughts, perceptions and feelings within an organization (Veldhoven, Vanthienen, 2021).

A cultural Digital Transformation change involves the introduction of new technologically advanced artifacts, with the consequent need to learn how to use them and to conceive professional roles. New values are required, in line with the evolution of the world in which organizations operate, increasing the focus on innovation, efficiency, costumer experience, creation of shared value and sustainability (Veldhoven, Vanthienen, 2021). Every cultural change involves resistance from organizational actors, who tend to take refuge in old habits driven by two main fears (Schein, 2000): being temporarily unable to work, workers and managers may in fact perceive that they do not possess the necessary skills to adapt to the new work, with the risk of being cut off, triggering the second fear, namely the exclusion (Fitzgerald et al., 2014).

Successful cultural change comes from managing these resistances, in integration with a clear vision of the direction to take. The challenges implied in the Digital Transformation scenario require workers to develop new skills such as soft skills, STEM skills and problem solving. HRM is called to think new ways of participation and of doing business, creating new products and services and providing new ways of organizing the workforce, all at a distance (Bondarouk and Brewster, 2016, p. 2662). The effect on competences (Ruël and Bondarouk, 2014; Bondarouk et al., 2015; Hecklau et al., 2016) highlights the need to develop both transversal skills (from execution to entrepreneurship) and digital capabilities for the interaction between humans and machines (Fregnan et al., 2020).

Impact on Workers

What are the main implications for workers triggered by the changes of Digital Transformation and Fourth Industrial Revolution? In addition to the above-mentioned challenges, typical of any cultural change, there are revolutions in daily life that must be faced.

Pais et al. (2021) present the main areas of change in the world of work of the digital age affecting the quality of work life. These are: time and space, complexity, autonomy, control and social relationships (Pais et al., 2021; Pais et al., 2020). In addition to these categories, we also mention health, safety and worker well-being.

Time and Space

The changes triggered by digital technologies on time and space in the workplace are afferent to this dimension, particularly with respect to issues of flexibility (and permeability), of the boundaries between personal and work life (Webster, 1997; Broadbent, 2015). Concerning time, one thinks of the intensification of rhythms or the discontinuity of working hours; with regard to space, the new ubiquity of technologies affects the symbolic dimension of places in determining modes of organization, forms of coordination and identity (Pais, et al, 2021). While on the one hand this opens up a space for greater flexibility in the management of one's work and the possibility to reconcile personal and work life, on the other hand there is a significant risk: these new time-space dimension creates conditions of exaggerated flexibility, fragmentation of work and pressing and overlapping deadlines (Pais et al., 2021).

Complexity: Knowledge and Skills, Tasks and Roles

Digital Transformation implies a shift in the content of the tasks of those working in 4.0 transformation contexts and their professional roles. In operations, a technical and specialized worker will be increasingly required also for digital and not only operational matters. Apparently, roles will be enriched with broader knowledge sets; on the other hand, new issues will emerge: (Pais et al., 2021; Pais et al., 2020)

- Digital coordination brings a number of extra activities, such as the need for information exchanges and their management. These activities require a significant amount of time and are not always perceived as enriching one's knowledge.

- There are effects on the new modes of coordination and interaction. The flow of information to be managed, the communication mediated by digital channels risks complicating the processes of exchange and management.

The increasing demands that emerge in terms of competencies concern skills such as adaptability, development of attitudes and social skills.

Autonomy and Control

The term "Datafication" refers to the use of technology to track individual choices (citizen or worker) and to transform social action into data. Digital Transformation has introduced into the world of work various technologies designed to collect data of different kinds, concerning schedules, production rhythms and even physiological indicators of the worker (Pais et al., 2021). The Digital Transformation has further thinned the boundary between control and autonomy in the world of work; if on the one hand the workers have greater freedom in carrying out their activities, on the other hand they are increasingly subject to constant control mechanisms related to the monitoring of their activities. In this regard, an ethical issue comes into play, linked not only to privacy but also, and above all, to issues of individual freedom. The role of employees will change significantly as increasingly real-time control will transform the content of work, processes and the work environment, and the implementation of a different socio-technical approach to work organization will offer workers the opportunity to enjoy greater responsibility and improve personal development (Pais et al., 2021).

Social Relations

The nature of social relations is significantly altered by Digital Transformation, new forms of human-machine and human-human interaction mediated by technology are observed. The digital spaces, used for informal communication and for work purposes, promote the exchange and sharing of information. The dynamics linked to the new interactive and relational modalities are still under investigation (Pais et al., 2021).

Health, Safety and Worker Well-Being

Regarding the impact on the world of work, it is stressed that Industry 4.0 offers rich opportunities for the improvement of health and safety conditions of workers. If an approach that is also focused on the health and safety needs of workers prevails, innovation 4.0 could bring significant benefits and improvements for everyone. If, on the other hand, a vision purely focused on economic aspects prevails, in a framework that is not adequately regulated, the risk of a negative impact on workers' health is instead possible in several areas, in particular (Tronci, 2018; Pais et al., 2021):

- intensification of workload,
- increased organizational constraints,
- information overload,
- depersonalization with loss of sense of belonging and attachment to work,
- difficulty in separating private and professional life.

From this point of view, it is inevitable that it is necessary to rethink and adapt the protection of workers with respect to new technologies and work environments, that will determine an evolution in terms of operating methods and related risks, skills and professionalism required.

Operational Efficiency

Although digital technologies can possibly change an organization, we discovered a day and a half featuring operational effectiveness – which incorporates the computerization (Andriole, 2017), the improvement of business measures (Gust et al., 2017) just as costs reserve funds (Pagani, 2013), as an advantage of Digital Transformation. This phennomenon is also associated with increases in several dimensions of organizational performance, including innovativeness (Svahn et al., 2017), financial performance (Karimi, Walter, 2015), firm growth (Tumbas et al., 2015), reputation (Kane, 2016; Yang et al., 2012) as well as competitive advantage (Neumeier et al., 2017). For example, under the freemium model, a firm can use online communities to increase the sense of belonging of users and motivate them to purchase premium accounts (Oestreicher- Singer and Zalmanson, 2012). With regards to innovative firms where the development rate is nonlinear, Tumbas et al. (2015) tracked down that competitive firms set up a "digital facade" to empower network with clients and colleagues, while later utilizing this exterior as a tool to cultivate associations with different clients and providers. These and different models (Setia et al., 2013) show how digital technologies can, through higher client commitment and cooperation, encourage higher benefits for firms. On a theoretical level, it has been suggested that digital technologies can sustain an organization's ability to detect the complexity of its context to plan a reaction that can help amplifying its odds of endurance through the transformation or the redefinition of its center exercises (Tanriverdi, Lim, 2017).

Digital Transformation and Customer Experience

Improving customer experience probably originates the main inspirations for organizations to carry out Digital Transformation (Kruschwitz et al., 2014). Digital Transformation can improve client administrations and upgrade customer experience (Venkatesh et al., 2013). It can help companies to assemble dynamic and adaptable plans of action for the creation of shared value (Knapp et al., 2017). It can improve data analysis abilities for a better comprehension of the customer (Bonnet et al., 2014). This deeper comprehension of the basic achievement variables of advanced change for improving customer experience would guarantee its effective execution in organizations. Digital Transformation also increases and improves communication with customer (Westerman, Bonnet, 2015). This prompts a huge improvement of the customer experience (Palmer et al., 2015; Ward and Peppard, 2016).

There are various business factors that can affect the success of Digital Transformation to improve the customer experience. These factors include value proposition (Berman, 2012), cost optimization (Sambamurthy and Zmud, 2012), market analysis (El Sawy et al., 2013), flexible pricing (Dong et al., 2006), product implementation strategy (Ward, Peppard, 2016), customer analysis (Hess et al., 2015), cost reduction (Cartwright et al., 2012), product demand and supply (Dong et al., 2006), regulations and policies (Gao et al., 2010). The consideration of these factors is significant towards the success of Digital Transformation to improve the customer experience. There are also multiple digital dimensions that can trigger the success of Digital Transformation for the improvement of the customer experience, including usability (Zhou, 2011), availability (Kruschwitz et al., 2014), infrastructure capacity (Palmer et al., 2015), data security (Dong et al., 2006) and digital content (Palmer et al., 2015). These factors play a crucial role in DT for the improvement of the customer experience in organizations (Sahu, Deng, Mollah, 2018; Sahu, 2019).

Digital Transformation and New Skills

We are facing a new way of approaching everyone's daily work, and it will most likely involve leaving the comfort zone and learning new skills even for those in the company who are directing this process of change (Sousa, Rocha, 2019). When an organization begins a Digital Transformation journey, it realizes it needs different skill sets rather than the traditional ones, in order to meet the demands generated by technological change (Werner, Wager, 2019). The above-mentioned development of new skills revolves around the abilities to be more agile, people-oriented, innovative, customer-focused, lean, efficient and able to induce/exploit opportunities to change the status quo quickly, generating revenue from the arising new information and services (Sousa, Rocha, 2019). With regards to Digital Transformation, changes to the design just as the way of life of an association lead workers to expect jobs that were generally outside of their abilities.

As digital technologies enable new forms of automation (Neumeier et al., 2017) and decision-making (Dremel et al., 2017; Hess et al., 2016), questions on the need to develop the skills of existing workers (Hess et al., 2016), as well as the skills required for future workers who will form the digital workforce (Colbert et al., 2016), are also becoming increasingly relevant (Watson, 2017). A long way from eliminating the requirement for organizations to depend on human resources, Digital Transformation expects workers improving their analytical skills to solve increasingly complex business problems (Dremel et al. 2017), and supporting workers through this change extends far beyond the domain of HR (Karimi and Walter, 2015; Singh and Hess, 2017).

The desk-based research highlights the nature of **Digital Transformation as a process where digital technologies create an impetus for organizations to implement responses to gain or maintain their competitive advantage** (Werner, Wager, 2019). But what do we mean when we talk about competencies, and what kind of nature should these competencies have to accompany the access into the 4.0 world of work (Sousa, Rocha, 2019)?

Competencies are one of the most studied constructs within the psychological, social and organizational sciences, there are quite distinct models for which, again, a choice of positioning is necessary (Sousa, Rocha,

2019). In this chapter authors considered competencies within a trichotomy model composed by the following elements:

- *Knowledge*: A body of information that a person possesses that may be applied directly to the performance of a function. It includes events, systems, facts, theories, methods, cases and concepts that results from formal education training or personal experience.
- *Skill*: Demonstrated and observable ability to perform a task with ease and proficiency. It often requires the use of equipment, machinery, tools or automated systems. It also implies measurable performance.
- *Attributes*: An individual's demonstrated personality traits. Attributes are often broader and more abstract than skills or knowledge, as they include personal characteristics like attitude, motivation, ambition, values and demeanor.

This model represents a comprehensive competency framework that provides organizations and its employees with a common understanding of the knowledge, skills, attributes and behaviors required to achieve organizational goals and remain competitive (Sousa, Rocha, 2019). The literature shows different types of competencies, authors have focused on those most suitable to adequately manage the scenario of Digital Transformation and company evolution. These have been grouped into two main categories: STEM skills and Capabilities.

STEM Skills

STEM stands for Science, Technology, Engineering and Mathematics. **STEM skills**, are **the individual skills needed for science, mathematics and engineering, as well as those needed to use technology effectively**. STEM skills include numeracy and the ability to generate, understand and analyze empirical data including critical analysis; an understanding of scientific and mathematical principles; the ability to apply a systematic and critical assessment of complex problems with an emphasis on solving them the ability to communicate scientific issues to stakeholders and others (Prinsley, Baranyai, 2013). Core STEM subjects typically include: Mathematics; Chemistry; Computer Science; Biology; Physics; Architecture; and General, Civil, Electrical, Electronics, Communications, Mechanical, and Chemical Engineering. Recent studies identify a shortage of STEM professionals and the need to engage students at all levels in science to boost the supply of STEM workers (Prinsley, Baranyai, 2013). There are emerging skills shortages, skills gaps and recruitment difficulties. These problems are likely to get worse as the demand for STEM skills increases due to the pivotal role that these skills and jobs play in the economic, social and organizational development (Prinsley, Baranyai, 2013).

A number of solutions have been developed to address the different facets of STEM skills problems. These include the **need to combine the STEM skills of graduates with the 'soft' employability skills** as communication skills, team working and creative thinking, which help applying STEM skills in the business world and which are important to innovation (Werner, Wager, 2019). More specifically, below are some of the **essential STEM skills** you will want to develop and/or enhance (Van Laar et al., 2017).

- *Problem Solving*: STEM problems require to quickly work to make sense of problems as they are presented, acting productively to propose solutions (Van Laar et al., 2017).
- *Creativity*: STEM requires the ability to propose solutions to a problem through multiple approaches. In STEM, mistakes and failed attempts are positive experiences, offering opportunities for deeper learning (Van Laar et al., 2017).
- *Inquiry skills*: STEM requires hands-on, active participation to solve problems effectively. Students are the drivers of solutions and should be asking questions, proposing ideas, generating and testing solutions, then making decisions based on data to understand how to further refine ideas (Van Laar et al., 2017).
- *Math and Science skills*: The mathematics and science skills are the foundation of STEM and must be applied in pursuit of solutions. (Van Laar et al., 2017).
- *Engineering-Design Thinking*: Thanks to this kind of thinking, it is possible to identify the problem, research solutions, build prototypes, test, redesign, test again and iterate further as needed. Each step moves closer to the creation of a solution (Van Laar et al., 2017).
- *Critical Thinking*: Effective STEM learning requires to analyze information, evaluate designs, reflect on thinking, synthesize new ideas, and propose creative solutions. (Van Laar et al., 2017).
- *Collaboration*: Big challenges are rarely solved by individuals. Working on STEM problems also involves learning to work as a part of a team (Van Laar et al., 2017).

Capabilities and Digital Transformation

There is a fit between capabilities and Digital Transformation (Van Laar et al., 2017). The literature highlights the nature of Digital Transformation as a source of disruptive change in a company's structure and competitive environment (Vial, 2019). Vial (2019, p.5) argues that "The contributions of capabilities have been found most useful in contexts fraught with environmental turbulence or hypercompetition, as ordinary competencies models cannot explain - on their own - how firms build and sustain competitive advantage (Teece, 2014)". Physical resources are becoming comparatively less relevant than services, as consumers contribute to influence trends related to the use of digital technologies, and as value networks become broader and more complex, firms experience higher levels of uncertainty. To manage this uncertainty, mechanisms to sense and adapt to changes that originate outside of the firm's competitive environment and locus of control must be put in place. Although we have some evidence of firms managing to adapt to these changes (Yeow et al., 2017), our understanding of the ability for firms to design repeatable mechanisms for this purpose is limited. "Developments in Digital Transformation literature have proposed the concept of digital maturity (Kane, 2017) as a capacity to respond to change in an appropriate manner, and we argue that Capabilities may help us understand how firms achieve digital maturity as they design and maintain these higher-level mechanisms enabling adaptability through successive waves of digital innovation (Van Laar et al., 2017)." (Vial 2019, p.8)

In particular, **a set of 7 capabilities** referable to as many dimensions of the current scenario has been identified (Pinto et al., 2021):

- *Deal with Human*: Living in a world so deeply influenced by technology, the human dimension plays an essential role for both companies having the intention to survive change and persons having to adapt to the surrounding reality. The human nature, along with its needs, must be put at the center (Romero et al. 2016).
- *Deal with Technology*: As a natural consequence of the digital revolution, today it is essential to be familiar with the new technologies and taking advantage of the opportunities they offer. It is an unavoidable requirement for organizations to implement them in their processes as well as it is for workers (Bersin, 2019).

- *Agility*: A new culture of work implies flexibility and open-mindedness; it is fundamental thinking out-of-the-box and to react quickly, in a creative process often including the client/consumer itself (Schwab, 2016).
- *Engagement*: being curious and proactive, properly balancing passion and rationality, Feeling highly involved; all of these elements allow to add something personal to each experience or activity or experience (Ostroff e Bowen, 2000).
- *Collaboration*: Complex activities require the cooperation between people, that represents the possibility to increase the effectiveness of a performance both quantitatively and qualitatively (Hecklau et al., 2017).
- *Interdisciplinarity*: The new work culture promotes the integration of different kinds of knowledge in order to broaden solutions and perspectives (Blinder, 2008).
- *Innovation*: unpredictable and fats changes are constantly produced by the new technologies, therefore it is indispensable to conceive innovation being always open to to the combination of data and information, the experimentation of different solutions and to the introduction of something new. Just imagining new ways can lead to the generation of value (Bondarouk and Brewster, 2016).

It was also possible to identify **a set of competences 4.0 or 21st century skills** defined by Comau (Fregnan et al.,2020):

- *Critical Thinking*: is an in-depth analysis of facts to form a judgment. Critical thinking as self-disciplined, self-corrective thinking,self-monitored and self-directed;
- *Cognitive flexibility*: It is the ability to overcome usual thoughts, behaviors and responses and thus adapting them to new situations. It is also the ability to adjust one's thinking, shifting from old situations to new ones;
- *Open-mindedness*: It means to be receptiveness to new ideas and concepts. It relates to the willingness to take new viewpoints seriously and the way in which people approach the knowledge of others;
- *Complex Problem Solving*: It means identifying complex problems and reviewing the related information for the implementation of solutions. It is the ability to solve new, ill-defined problems in a complex, real-world setting;

- *Creativity*: it is the process of becoming sensitive to gaps in knowledge, problems, deficiencies, missing elements, and so on; it means identifying the difficulty and searching for solutions, formulating hypotheses about the deficiencies, testing and retesting such hypotheses and finally communicating the results;
- *Cooperative Networking*: it is expertise and knowledge sharing. independency and autonomy are combined with the advantage of learning from others and being able to modify the way one thinks, behaves and works;
- *Social Intelligence*: it is the ability to relate to others in an efficient and socially compatible way;
- *Emotional Intelligence*: it is the capability of individuals to recognize their own emotions and those belonging to others, to use the emotional knowledge to guide thinking and behavior, to manage and/or adjust emotions adapting to environments or to achieve one's goals;
- *Judgment*: it is the ability to come to sensible conclusion or make reasonable decisions.
- *Decision Making*: the process of selecting a logical choice from the available options, considering all the options and weighting the negative and positive aspects of each alternative.
- *Influence*: it is the ability to produce a significant effect on someone. If someone influences someone else, they are changing a person in an indirect but relevant way.

All these skills are crucial not only for an **adequate management of the innovations** implied by the Digital Transformation, but also and above all to **anticipate the demands of tomorrow's work**. In this sense, **initiatives aimed at developing skills for jobs with a high digital content represent a way of creating and sharing social value**.

CONCLUSION

The desk-based research has allowed to understand the technological revolution we are witnessing in terms of Fourth Industrial Revolution, Industry 4.0 and Digital Transformation. This scenario has considerable implications at both social and organizational level, as it significantly modifies the way we interact with each other, activities and work processes, hierarchical structures and organizational culture. On the other hand, Digital Transformation presents great development opportunities for business and for sharing value with the society, in a relationship of mutual exchange with the external stakeholders of a company.

It is indeed true that the Digital Transformation scenario and the company evolution must be supported by the **development of new skills** (STEM and Capabilities) **to master the new work processes and contents, as well as the new modes of interaction**. More in detail, the authors have identified 7 Capabilities, already deepened in a previous research, crucial for the workers of Industry 4.0 and for the workers of tomorrow: Deal with Human; Deal with Technology; Agility; Engagement; Collaboration; Interdishiplinarity; Innovation. They also have identified a set of competences 4.0 or 21st century skills: Critical Thinking; Cognitive Flexibility; Open-mindedness; Complex Problem Solving; Creativity; Cooperative Networking; Social Intelligence; Emotional Intelligence; Judgment; Decision Making; Influence. The development of such skills can be sustained by means of **educational programs** that place **new technologies** (in particular robotic assistants) at the center, in order to facilitate the entry into the 4.0 world of work.

In the next chapter authors will present a particular way of applying Digital Transformation for educational purposes, to be understood as a generative and value-creating action: the **Educational Robotics initiatives.**

REFERENCES

Agarwal, R., Gao, G., DesRoches, C., & Jha, A. K. (2010). Research commentary—The digital transformation of healthcare: Current status and the road ahead. *Information Systems Research*, *21*(4), 796–809. doi:10.1287/isre.1100.0327

Amhøj, C. B. (2019). Sustainability as an ecology of learning, thinking and acting: Using the World Health Organization's six P's as an action-research intervention to create public value with multiple bottom lines. *Local Economy*, *34*(5), 439–455. doi:10.1177/0269094219847124

Andrews, G., & Russell, M. (2012). Employability skills development: strategy, evaluation and impact. *Higher Education, Skills and Work-Based Learning*, *2*(1).

Andriole, S. J. (2017). Five myths about digital transformation. *MIT Sloan Management Review*, *58*(3).

Berghaus, S., & Back, A. (2016, September). Stages in Digital Business Transformation: Results of an Empirical Maturity Study. MCIS, 22.

Berman, S. J. (2012). Digital transformation: Opportunities to create new business models. *Strategy and Leadership*, *40*(2), 16–24. doi:10.1108/10878571211209314

Bersin, J., & Chamorro-Premuzic, T. (2019). New ways to gauge talent and potential. *MIT Sloan Management Review*, *60*(2), 1.

Bharadwaj, A., El Sawy, O. A., Pavlou, P. A., & Venkatraman, N. V. (2013). Digital business strategy: Toward a next generation of insights. *Management Information Systems Quarterly*, *37*(2), 471–482. doi:10.25300/MISQ/2013/37:2.3

Blinder, A. S. (2008). *The quiet revolution*. Yale University Press.

Bondarouk, T., & Brewster, C. (2016). Conceptualising the future of HRM and technology research. *International Journal of Human Resource Management*, *27*(21), 2652–2671. doi:10.1080/09585192.2016.1232296

Bondarouk, T., Parry, E., & Furtmueller, E. (2017). Electronic HRM: Four decades of research on adoption and consequences. *International Journal of Human Resource Management*, *28*(1), 98–131. doi:10.1080/09585192.2016.1245672

Broadbent, J., & Poon, W. L. (2015). Self-regulated learning strategies & academic achievement in online higher education learning environments: A systematic review. *The Internet and Higher Education*, *27*, 1–13. doi:10.1016/j.iheduc.2015.04.007

Bughin, J., Deakin, J., & O'Beirne, B. (2019). Digital transformation: Improving the odds of success. *The McKinsey Quarterly*.

Carter, D. (2018). How real is the impact of artificial intelligence? The business information survey 2018. *Business Information Review*, *35*(3), 99–115. doi:10.1177/0266382118790150

Chanias, S. (2017). Mastering digital transformation: the path of a financial services provider towards a digital transformation strategy. *Proceedings of the 25th European Conference on Information Systems (ECIS)*.

Colbert, A., Yee, N., & George, G. (2016). The digital workforce and the workplace of the future. *Academy Journal of Management*, *54*(3), 731–738. doi:10.5465/amj.2016.4003

Dremel, C., Wulf, J., Herterich, M. M., Waizmann, J. C., & Brenner, W. (2017). How AUDI AG Established Big Data Analytics in Its Digital Transformation. *MIS Quarterly Executive*, *16*(2).

Earley, M. A. (2014). A synthesis of the literature on research methods education. *Teaching in Higher Education*, *19*(3), 242–253. doi:10.1080/13562517.2013.860105

Ebert, C., & Duarte, C. H. C. (2018). Digital Transformation. *IEEE Software*, *35*(4), 16–21. doi:10.1109/MS.2018.2801537

El Sawy, O. A., & Pereira, F. (2013). Digital business models: review and synthesis. *Business Modelling in the Dynamic Digital Space*, 13-20.

Fenech, R., Baguant, P., & Ivanov, D. (2019). The changing role of human resource management in an era of Digital Transformation. *Journal of Management Information and Decision Sciences*, *22*(2), 1–10.

Fitzgerald, M., Kruschwitz, N., Bonnet, D., & Welch, M. (2014). Embracing digital technology: A new strategic imperative. *MIT Sloan Management Review*, *55*(2), 1.

Frank, A. G., Mendes, G. H., Ayala, N. F., & Ghezzi, A. (2019). Servitization and Industry 4.0 convergence in the digital transformation of product firms: A business model innovation perspective. *Technological Forecasting and Social Change*, *141*, 341–351. doi:10.1016/j.techfore.2019.01.014

Fregnan, E., Ivaldi, S., & Scaratti, G. (2020). HRM 4.0 and New Managerial Competences Profile: The COMAU Case. *Frontiers in Psychology*, *11*, 11. doi:10.3389/fpsyg.2020.578251 PMID:33329228

Gao, S., Hakanen, E., & Rajala, R. (2020, January). Digital transformation: The interplay of explorative and exploitative capability development. *Proceedings of the 53rd Hawaii International Conference on System Sciences*. 10.24251/HICSS.2020.527

Grubbs, M. (2013). Robotics intrigue middle school students and build STEM skills. *Technology and Engineering Teacher, 72*(6), 12.

Guerin, B., Janta, B., van Gorp, A. (2018). Desk-based research and literature review. *Evaluating interventions that prevent or counter violent extremism*, 63.

Gust, G., Neumann, D., Flath, C. M., Brandt, T., & Ströhle, P. (2017). How a traditional company seeded new analytics capabilities. *MIS Quarterly Executive*, *16*(3), 215–230.

Hanelt, A., Piccinini, E., Gregory, R. W., Hildebrandt, B., & Kolbe, L. M. (2015). Digital Transformation of Primarily Physical Industries-Exploring the Impact of Digital Trends on Business Models of Automobile Manufacturers. *Wirtschaftsinformatik*, *88*(3-5), 1313–1327.

Hartl, E., & Hess, T. (2017). *The role of cultural values for digital transformation: Insights from a Delphi study*. Academic Press.

He, Q., Meadows, M., Angwin, D., Gomes, E., & Child, J. (2020). Strategic alliance research in the era of digital transformation: Perspectives on future research. *British Journal of Management*, *31*(3), 589–617. doi:10.1111/1467-8551.12406

Heavin, C., & Power, D. J. (2018). Challenges for digital transformation–towards a conceptual decision support guide for managers. *Journal of Decision Systems, 27*(sup1), 38-45.

Hecklau, F., Galeitzke, M., Flachs, S., & Kohl, H. (2016). Holistic approach for human resource management in Industry 4.0. *Procedia CIRP*, *54*, 1–6. doi:10.1016/j.procir.2016.05.102

Henriette, E., Feki, M., & Boughzala, I. (2015). The shape of digital transformation: a systematic literature review. *MCIS 2015 Proceedings, 10*, 431-443.

Hess & Benlian. (2015). Digital Transformation strategies. *Business & Information Systems Engineering, 57(* 5) , 339-343. doi:10.100712599-015-0401-5

Hess, M. C. T., Benlian, A., & Wiesböck, F. (2016). Options for formulating a digital transformation strategy. *MIS Quarterly Executive*, *15*(2).

Hinterhuber, A., Vescovi, T., & Checchinato, F. (Eds.). (2021). *Managing Digital Transformation: Understanding the Strategic Process*. Routledge. doi:10.4324/9781003008637

Hoberg, P., Krcmar, H., & Welz, B. (2017). *Skills for digital transformation.* Technical University of Munich, Chair for Information Systems, Study.

Ismail, M. H., Khater, M., & Zaki, M. (2017). Digital business transformation and strategy: What do we know so far. *Cambridge Service Alliance*, *10*, 1–35.

Kane, G. C. (2016). Crowd-based capitalism? Empowering entrepreneurs in the sharing economy. *MIT Sloan Management Review*, *57*(3).

Kane, G. C., Palmer, D., Nguyen-Phillips, A., Kiron, D., & Buckley, N. (2017). Achieving digital maturity. *MIT Sloan Management Review*, *59*(1).

Kane, G. C., Palmer, D., Phillips, A. N., & Kiron, D. (2015a). Is your business ready for a digital future? *MIT Sloan Management Review*, *56*(4), 37.

Kane, G. C., Palmer, D., Phillips, A. N., Kiron, D., & Buckley, N. (2015). Strategy, not technology, drives digital transformation. *MIT Sloan Management Review and Deloitte University Press*, *14*, 1–25.

Karimi, J., & Walter, Z. (2015). The role of dynamic capabilities in responding to digital disruption: A factor-based study of the newspaper industry. *Journal of Management Information Systems*, *32*(1), 39–81. doi:10.1080/07421222.2015.1029380

Kaufman, I., & Horton, C. (2014). *Digital marketing: Integrating strategy and tactics with values, a guidebook for executives, managers, and students*. Routledge. doi:10.4324/9781315879451

Knapp, J. A., Gearhart, A., Kellerman, L. S., & Klimczyk, L. (2018). The Pennsylvania Newspaper Archive: Harnessing an open-source platform to host digitized collections online. *IFLA Journal, 44*(2), 143–153.

Lee, M. X., Lee, Y. C., & Chou, C. J. (2017). Essential implications of the digital transformation in industry 4.0. *Journal of Scientific and Industrial Research*, *76*, 465–467.

Legner, C., Eymann, T., Hess, T., Matt, C., Böhmann, T., Drews, P., Mädche, A., Urbach, N., & Ahlemann, F. (2017). Digitalization: Opportunity and challenge for the business and information systems engineering community. *Business & Information Systems Engineering*, *59*(4), 301–308. doi:10.100712599-017-0484-2

Li, G., Hou, Y., & Wu, A. (2017). Fourth Industrial Revolution: Technological drivers, impacts and coping methods. *Chinese Geographical Science*, *27*(4), 626–637. doi:10.100711769-017-0890-x

Li, G., Hou, Y., & Wu, A. (2017). Fourth Industrial Revolution: Technological drivers, impacts and coping methods. *Chinese Geographical Science*, *27*(4), 626–637. doi:10.100711769-017-0890-x

Loebbecke, C., & Picot, A. (2015). Reflections on societal and business model transformation arising from digitization and big data analytics: A research agenda. *The Journal of Strategic Information Systems*, *24*(3), 149–157. doi:10.1016/j.jsis.2015.08.002

Lu, B., & Zl, M. (2017). Digital transformation disruptive threats identification. *Industry 4.0, 2*(2), 64-65.

Matt, C., Hess, T., & Benlian, A. (2015). Digital transformation strategies. *Business & Information Systems Engineering*, *57*(5), 339–343. doi:10.100712599-015-0401-5

McKeown, I., & Philip, G. (2003). Business transformation, information technology and competitive strategies: Learning to fly. *International Journal of Information Management*, *23*(1), 3–24. doi:10.1016/S0268-4012(02)00065-8

Morrar, R., Arman, H., & Mousa, S. (2017). The fourth industrial revolution (Industry 4.0): A social innovation perspective. *Technology Innovation Management Review*, *7*(11), 12–20. doi:10.22215/timreview/1117

Nalajala, T. D. (2003). *Internet telerobotic system*. University of Nevada.

Neumeier, A., Wolf, T., & Oesterle, S. (2017). *The manifold fruits of digitalization-determining the literal value behind*. Academic Press.

Newell, S., & Marabelli, M. (2015). Strategic opportunities (and challenges) of algorithmic decision-making: A call for action on the long-term societal effects of 'datification'. *The Journal of Strategic Information Systems*, *24*(1), 3–14. doi:10.1016/j.jsis.2015.02.001

Oestreicher-Singer, G., & Zalmanson, L. (2013). Content or community? A digital business strategy for content providers in the social age. *Management Information Systems Quarterly*, *37*(2), 591–616. doi:10.25300/MISQ/2013/37.2.12

Pagani, M. (2013). Digital business strategy and value creation: Framing the dynamic cycle of control points. *Management Information Systems Quarterly*, *37*(2), 617–632. doi:10.25300/MISQ/2013/37.2.13

Pais, I., Arcidiacono, D. L., Samuele, P., & Rosina, A. (2020). *Competenze digitali e divari regionali nelle PMI italiane*. Academic Press.

Pais, I., Ponzellini, A. M., Salvatore, C., & Della Rocca, G. (2021). *La qualità del lavoro nella trasformazione digitale*. Academic Press.

Pais, I., & Provasi, G. (2015). Sharing economy: A step towards the re-embeddedness of the economy? *Stato e mercato, 35*(3), 347-378.

Patel, M. B., McCarthy, T. J., Morris, P. W. G., & Elhag, T. M. S. (2000). The role of IT in capturing and managing knowledge for organisational learning on construction projects. *Proceedings of CIT*, 674-85.

Peppard, J., & Ward, J. (2016). *The strategic management of information systems: Building a digital strategy*. John Wiley & Sons.

Pereira, A. C., & Romero, F. (2017). A review of the meanings and the implications of the Industry 4.0 concept. *Procedia Manufacturing*, *13*, 1206–1214. doi:10.1016/j.promfg.2017.09.032

Philbeck, T., & Davis, N. (2018). The fourth industrial revolution. *Journal of International Affairs*, *72*(1), 17–22.

Piccinini, E., Gregory, R. W., & Kolbe, L. M. (2015). Changes in the producer-consumer relationship-towards digital transformation. *Changes (Hove, England)*, *3*(4), 1634–1648.

Piccinini, E., Hanelt, A., Gregory, R., & Kolbe, L. (2015). *Transforming industrial business: The impact of digital transformation on automotive organizations*. Academic Press.

Pinto, D., Scaratti, G., & Fregnan, E. (2021). *Learning Agility 4.0: ecosistemi e trasformazione culturale*. Pearson.

Popper, K.R. (1994). *La conoscenza e il problema corpo mente*. Il mulino.

Porter, M. E., & Heppelmann, J. E. (2014). How smart, connected products are transforming competition. *Harvard Business Review*, *92*(11), 64–88.

Pramanik, P. K. D., Pal, S., & Choudhury, P. (2018). Beyond automation: the cognitive IoT. artificial intelligence brings sense to the Internet of Things. In Cognitive Computing for Big Data Systems Over IoT (pp. 1-37). Springer.

Prinsley, R. T., & Baranyai, K. (2013). *STEM skills in the workforce: what do employers want?* Office of the Chief Scientist.

Reddy, S. K., & Reinartz, W. (2017). Digital transformation and value creation: Sea change ahead. *GfK Marketing Intelligence Review, 9*(1).

Reis, J., Amorim, M., Melão, N., & Matos, P. (2018, March). Digital transformation: a literature review and guidelines for future research. In *World conference on information systems and technologies* (pp. 411-421). Springer. 10.1007/978-3-319-77703-0_41

Rocha, C. S., Antunes, P., & Partidário, P. (2019). Design for sustainability models: A multiperspective review. *Journal of Cleaner Production, 234*, 1428–1445. doi:10.1016/j.jclepro.2019.06.108

Romero, D., Gaiardelli, P., Powell, D., Wuest, T., & Thürer, M. (2018). Digital Lean cyber-physical production systems: the emergence of digital Lean manufacturing and the significance of digital waste. In I. Moon, G. M. Lee, J. Park, D. Kiritsis, & G. von Cieminski (Eds.), *APMS 2018. IAICT* (Vol. 535, pp. 11–20). Springer. doi:10.1007/978-3-319-99704-9_2

Ruël, H., & Bondarouk, T. (2014). E-HRM research and practice: facing the challenges ahead. In *Handbook of strategic e-Business management* (pp. 633–653). Springer. doi:10.1007/978-3-642-39747-9_26

Rüßmann, M., Lorenz, M., Gerbert, P., Waldner, M., Justus, J., Engel, P., & Harnisch, M. (2015). *Industry 4.0: The future of productivity and growth in manufacturing industries.* Boston Consulting Group.

Sahu, N. (2019). *Investigating the Critical Success Factors of Digital Transformation for Improving the Customer Experience in Australian Organisations* (Doctoral dissertation). RMIT University, Melbourne, Australia.

Sahu, N., Deng, H., & Mollah, A. (2018). Investigating the critical success factors of Digital Transformation for improving customer experience. In *International Conference on Information Resources Management (CONF-IRM).* Association For Information Systems.

Sambamurthy, V., & Zmud, R. W. (2012). *Guiding the digital transformation of organizations.* Legerity Digital Press.

Schein, E. H. (1999). The Corporate culture survival guide. Jossey-Bass.

Schuchmann, D., & Seufert, S. (2015). Corporate learning in times of digital transformation: A conceptual framework and service portfolio for the learning function in banking organisations. *International Journal of Corporate Learning, 8*(1), 31–39. doi:10.3991/ijac.v8i1.4440

Schwab, K. (2015). *World economic forum. Global Competitiveness Report (2014-2015).* Retrieved http://www3. weforum. org/docs/WEF_GlobalCompetitivenessReport_2014-15. pdf

Schwab, K. (2016). *La quarta rivoluzione industriale*. FrancoAngeli.

Schwab, K. (2017). *The fourth industrial revolution*. Currency.

Schwab, K., & Davis, N. (2018). *Shaping the future of the fourth industrial revolution*. Currency.

Setia, P., Setia, P., Venkatesh, V., & Joglekar, S. (2013). Leveraging digital technologies: How information quality leads to localized capabilities and customer service performance. *Management Information Systems Quarterly*, *37*(2), 565–590. doi:10.25300/MISQ/2013/37.2.11

Shahroom, A. A., & Hussin, N. (2018). Industrial revolution 4.0 and education. *International Journal of Academic Research in Business & Social Sciences*, *8*(9), 314–319. doi:10.6007/IJARBSS/v8-i9/4593

Singh, A., & Hess, T. (2017). How chief digital officers promote the digital transformation of their companies. *MIS Quarterly Executive*, *16*(1).

Srivastava, S. C., & Shainesh, G. (2015). Bridging the service divide through digitally enabled service innovations. *Management Information Systems Quarterly*, *39*(1), 245–268. doi:10.25300/MISQ/2015/39.1.11

Stolterman, E., & Fors, A. C. (2004). Information technology and the good life. In *Information systems research* (pp. 687–692). Springer. doi:10.1007/1-4020-8095-6_45

Subramaniam, M., Iyer, B., & Venkatraman, V. (2019). Competing in digital ecosystems. *Business Horizons*, *62*(1), 83–94. doi:10.1016/j.bushor.2018.08.013

Svahn, F., Mathiassen, L., Lindgren, R., & Kane, G. C. (2017). Mastering the digital innovation challenge. *MIT Sloan Management Review*, *58*(3), 14.

Tanriverdi, H., & Lim, S.-Y. (2017). How to survive and thrive in complex, hypercompetitive, and disruptive ecosystems? The roles of IS-enabled capabilities. *International Conference of Information Systems*, Seoul, South Korea.

Teece, D. J. (2014). The foundations of enterprise performance: Dynamic and ordinary capabilities in an (economic) theory of firms. *The Academy of Management Perspectives*, *28*(4), 328–352. doi:10.5465/amp.2013.0116

Teece, D. J. (2014a). A dynamic capabilities-based entrepreneurial theory of the multinational enterprise. *Journal of International Business Studies*, *45*(1), 8–37. doi:10.1057/jibs.2013.54

Tumbas, S., Berente, N., Seidel, S., & vom Brocke, J. (2015). *The 'digital façade' of rapidly growing entrepreneurial organizations*. International Conference ofInformation Systems, Forth Worth, TX.

Van Laar, E., Van Deursen, A. J., Van Dijk, J. A., & De Haan, J. (2017). The relation between 21st-century skills and digital skills: A systematic literature review. *Computers in Human Behavior*, *72*, 577–588. doi:10.1016/j.chb.2017.03.010

Van Veldhoven, Z., & Vanthienen, J. (2021). Digital transformation as an interaction-driven perspective between business, society, and technology. *Electronic Markets*, 1–16. doi:10.100712525-021-00464-5

Venkatesh, V., Brown, S. A., & Bala, H. (2013). Bridging the qualitative-quantitative divide: Guidelines for conducting mixed methods research in information systems. *Management Information Systems Quarterly*, *37*(1), 21–54. doi:10.25300/MISQ/2013/37.1.02

Venkatraman, N. (1994). IT-enabled business transformation: From automation to business scope redefinition. *Sloan Management Review*, *35*, 73–73.

Verhoef, P. C., & Bijmolt, T. H. (2019). *Marketing perspectives on digital business models: A framework and overview of the special issue*. Academic Press.

Verhoef, P. C., Broekhuizen, T., Bart, Y., Bhattacharya, A., Dong, J. Q., Fabian, N., & Haenlein, M. (2021). Digital transformation: A multidisciplinary reflection and research agenda. *Journal of Business Research*, *122*, 889–901. doi:10.1016/j.jbusres.2019.09.022

Vey, K., Fandel-Meyer, T., Zipp, J. S., & Schneider, C. (2017). Learning & Development in Times of Digital Transformation: Facilitating a Culture of Change and Innovation. *International Journal of Advanced Corporate Learning*, *10*(1), 22. doi:10.3991/ijac.v10i1.6334

Vial, G. (2019). Understanding digital transformation: A review and a research agenda. *The Journal of Strategic Information Systems*, *28*(2), 118–144. doi:10.1016/j.jsis.2019.01.003

von Leipzig, T., Gamp, M., Manz, D., Schöttle, K., Ohlhausen, P., Oosthuizen, G., Palm, D., & von Leipzig, K. (2017). Initialising customer-orientated digital transformation in enterprises. *Procedia Manufacturing*, *8*, 517–524. doi:10.1016/j.promfg.2017.02.066

Wade, M. (2015). *Digital Transformation: A Conceptual Framework*. Global Center for Digital Business Transformation.

Warner, K. S., & Wäger, M. (2019). Building dynamic capabilities for digital transformation: An ongoing process of strategic renewal. *Long Range Planning*, *52*(3), 326–349. doi:10.1016/j.lrp.2018.12.001

Westerman, G., & Bonnet, D. (2015). Revamping your business through digital transformation. *MIT Sloan Management Review*, *56*(3), 10.

Westerman, G., Bonnet, D., & McAfee, A. (2014). *Leading digital: Turning technology into business transformation*. Harvard Business Press.

Yang, J. C., Chien, K. H., & Liu, T. C. (2012). A Digital Game-Based Learning System for Energy Education: An Energy COnservation PET. *Turkish Online Journal of Educational Technology-TOJET*, *11*(2), 27–37.

Yeow, A., Soh, C., & Hansen, R. (2018). Aligning with new digital strategy: A dynamic capabilities approach. *The Journal of Strategic Information Systems*, *27*(1), 43–58. doi:10.1016/j.jsis.2017.09.001

Yusuf, B., Walters, L. M., & Sailin, S. N. (2020). Restructuring Educational Institutions for Growth in the Fourth Industrial Revolution (4IR): A Systematic Review. *International Journal of Emerging Technologies in Learning*, *15*(3), 93. doi:10.3991/ijet.v15i03.11849

Zhou, Q. (2011). The Analysis of Digital Transformation of the Traditional Media Development in New Media Time. *China Media Report Overseas, 7*(4).

Zhu, K., Dong, S., Xu, S. X., & Kraemer, K. L. (2006). Innovation diffusion in global contexts: Determinants of post-adoption digital transformation of European companies. *European Journal of Information Systems*, *15*(6), 601–616. doi:10.1057/palgrave.ejis.3000650

Zinder, E., & Yunatova, I. (2016). Synergy for digital transformation: Person's multiple roles and subject domains integration. In *International Conference on Digital Transformation and Global Society* (pp. 155-168). Springer. 10.1007/978-3-319-49700-6_16

Chapter 3
Educational Robotics

ABSTRACT

In recent years we are witnessing an increasing diffusion of new technologies in school contexts, an aspect that implies changes in educational paradigms. The introduction of social agents such as robots in schools represents a significant opportunity both to innovate teaching and to respond to the needs introduced by the digital transformation and the Fourth Industrial Revolution. The use of social robots, which can be of diverse nature and form, is generally aimed at teaching curricular subjects (from STEM to humanities) or the principles of robotics. From an initial application in the field of special education, robots have begun to present themselves as valuable learning agents, capable of supporting high levels of student engagement, stimulating problem solving, teamworking, and communication skills, as well as supporting curricular learning.

INTRODUCTION

Education is an application area of Digital Transformation that contains comprehensive technology, personnel, and budget strategy (Barr, 2019). In the education field, common transformation efforts range from content conversion to automating administrative processes and integrating systems. As new learning applications, media channels, forms, as well as communication and collaboration technologies emerge, IT teams are faced with a re-evaluation of security, personnel, budget, and operational issues (Barr, 2019; Sundqvist, Koch, Birberg, Thornberg, Barr, Heimann, 2021). There are new meanings,

DOI: 10.4018/978-1-7998-8653-2.ch003

opportunities, and challenges in growing up in the reality with an increasing presence of Internet, where the role of robots and virtual assistants is constantly growing. The Fourth Industrial Revolution changes generate great implications in education and growth (Sundqvist et al., 2021). New technologies expand our relational resources, but at the same time they expose us to unprecedented risks, particularly significant for younger people who are still consolidating their identities. Among both scholars and practitioners there is a great interest in educational and socio-relational implications in human interactions mediated by virtual and robotic assistants (Barr, 2019).

This chapter aims to show how **the development of innovative educational experiences supported by robots could represent a way to create shared value with the community** (Sundqvist et al., 2021). But why is it so important to think about the link between Digital Transformation and education?

People spend many hours online every day, furthermore the Covid-19 pandemic has even increased the continuous connection between humans and computers (Barr, 2019). Today, a 5-year-old child, who cannot write yet, already has at least 500 hours of experienced technology (Sundqvist et al., 2021). At 18 years old, on the other hand, 15,000 hours are estimated, the time required to become an expert concert pianist. What happens to a child's brain when it is exposed to technologies for too long? There is a negative impact on language abilities, caused by an alteration in the white matter of the brain (Sundqvist et al., 2021). Technology impacts on white matter integrity in brain areas related to executive and linguistic functions (Barr, 2019). An early access to technology could be risky for a child, therefore it is functional to limit its use to 30 minutes a day until he or she is 6 years old, and using the devices to symbolically reproduce daily life, rather than just as a simple screen. A **responsible use of technology** aims to **acquire relational and social rules** that facilitate the access to society and relations (Sundqvist et al., 2021). It is indeed true that the use of technologies impacts on social and emotional intelligence too, for this reason we have to use it carefully.

But what is the positive aspect? Young people learn to use technology in a more intuitive way: older generations invest energy in learning and using technologies, while digital natives understand immediately what to do. Many technological companies, in fact, are launched by young people (Sundqvist et al., 2021). Moreover, digital natives make use of technologies not accessible to previous generations. Finally, they are able to incorporate technology as they use it; for example, while they are texting, they use more fingers than their parents (Barr, 2019). Technology also makes it possible to leave physical places and enter the digital world, a relational space that cannot be seen. There is a lack of body, of oneself and of the other; this could represent a limitation for the development of empathy and emotional intelligence but, on the other hand, the lack of the physical sphere can be an opportunity to win the shyness. The lack of contextual constraints helps to be more uninhibited (with more freedom of expression in both a positive and negative way), it is a resource to express themselves with greater freedom and has effects on well-being (Sundqvist et al., 2021). Through new technologies it is possible to develop weak forms of interaction, a relational opportunity. How to promote the positive use of new technologies? We need an **educational work** that considers the growth phase, focused on a number of key points (Sundqvist et al., 2021):

- Accompanying awareness (regulating emotions, empathy, skills) in the use of new technologies.
- Digital awareness: what risks and opportunities, what does it mean to use them?
- Working on rules and sharing.
- Providing positive role models (adults are the first ones to make an immoderate and unconscious use of technology and social media).

The chapter aims to **describe the field of Educational Robotics** on a social and organizational level, and then to **identify the main implications of the use of robots in education**. Since this is still an evolving phenomenon, authors have decided to describe it from a theoretical point of view, in order to make a positioning choice and to establish a knowledge starting point. In adherence with these goals and their nature, they identified **desk-based research as the most functional research method** (Guerin, Janta, Van Gorp, 2018).

This secondary research approach does not involve the collection of empirical material from the field, but rather the study of existing literature, with the aim of identifying the **theoretical anchors to understand and describe our phenomenon of interest**, namely the **Educational Robotics Field**, the kinds of robot that can be found in the educational field, the purposes with which they are used, and the main advantages linked to their application. Initially, the data collection has included an in-depth literature review of the concepts of "Educational Robotic", "Robotic Education" and "Educational

Robots". This step has allowed to identify **recurring elements for data processing**, which was accomplished by defining a set of categories and keywords as reference. The research phenomenon has been explored through the **constructs of Educational Robotics and its implications on learning outputs**. These elements will be presented in the next pages.

DATA COLLECTION AND PROCESSING

Social Robots in Educational Contexts

We have been hearing about robots for a long time, especially in the industrial context where they are used to improve efficiency and productivity. Nowadays, nonetheless, service robots are outnumbering industrial robots pointing out that robots are becoming more prevalent in society (Mubin, Stevens, Shahid, Mahmud, Dong, 2013). The concept of "Robot", as well as its physical and technical characteristics, has changed considerably over the years. Initially, robotics was mainly concerned with modeling, or the engineering and computer science aspects at the basis of the creation of the robot itself, considered as an artifact, generally distant from everyday reality. The growing applications in industry and the recent diffusion in the domestic field have shifted the attention of robotics also to the social level, in particular in relation to possible interventions of a psychological and pedagogical nature. These changes have led to a paradigmatic innovation, to a new conception of "robot", no longer seen as a "machine on wheels", but as a real organism and a social agent.

In fact, Social Robots have several characteristics, such as physicality, sensoriality or the need for food, that allow us to interact with them and perceive them as accessible. They also have motor systems, and they act physically in the world; this makes perceive them as organisms. In this perspective, the robot is considered as an organism because it has more or less complex sensory systems and it is able to support a communicative interaction. Finally, Robots need electric energy for their operation, they "eat" like us and this is what bring us to humanize them, they have a control system (algorithms, coding, artificial neural networks, etc.) that makes them autonomous in their operation.

The choice of positioning made in this chapter leads to consider the **robot as a social agent capable of introducing significant innovations in educational contexts**. A Social Robot is an autonomous or semi-autonomous robot capable of interacting and communicating with humans or other autonomous physical agents by following social behaviors and rules related to its specific role, as defined by Dautenhahn and Billard (1997): "embodied agents who have the capacity to recognize and engage in social interactions, perceive and interpret the world based on their own experience, and communicate and learn explicitly from one another".

The use of robots in education is not a novelty, Papert with his LOGO turtles was one of the field pioneers in the 1980s, but it wasn't until the advent of the Fourth Industrial Revolution that steps toward in introducing robots into the learning field became visible (Johal, Castellano, Tanaka, Okita, 2018). According to Rosenberg-Kima, Koren and Gordon (2020), in the last decade Social Robots have progressed drastically, this is particularly true in the field of education (Mubin et al., 2013; Brown and Howard, 2014; Gordon et al., 2015; Belpaeme et al., 2018b). Compared to some of their competitor, namely tablets and screens, "Social Robots have been shown to convey more learning gains (Wainer et al., 2006; Leyzberg et al., 2012; Li, 2015; Luria et al., 2017) and to evoke more emotional expressions (Spaulding et al., 2016)" (Rosenberg-Kima et al., 2020, p. 2). Social Robots have often been used to teach math (Brown and Howard, 2014 →Brown, L. N., & Howard, A. M. (2014, March) and science (Shiomi et al., 2015). Moreover, Social Robots have been used to develop meta-cognitive skills (Rosenberg-Kima et al., 2020) such as growth mindset (Park et al., 2017) and curiosity (Gordon et al., 2015; Ceha et al., 2019). Rosenberg-Kima, Koren and Gordon (2020) argue that a single robot can interact with small groups of children (Leite et al., 2015; Strohkorb et al., 2015), elderly (Matsuyama et al., 2008), and adults (Matsuyama et al., 2015). More specifically, according to the authors, (Rosenberg-Kima et al. 2020) several studies examined possible roles of Social Robots in group interaction (Jung et al., 2015; Shen et al., 2018; Alves-Oliveira et al., 2019; Correia et al., 2019; Oliveira et al., 2019). In particular, the educational contexts in which Social Robots are most used concern school, especially education, and just recently and in a still implementable way, the world of work.

If, on the one hand, the increasing diffusion of Social Robots with educational purposes can be explained in relation to the technological progress generated by Digital Transformation, which has brought significant improvements at both the software and hardware levels, on the other hand, the application of robots in educational contexts seems to respond to the need to promote greater familiarization with these tools, imagining that their use, along with the other elements of the digital revolution, will be increasingly consistent in our society.

Pedagogical Foundations of Educational Robotics: Piaget's Constructivism and Papert's Constructionism

The two most popular pedagogical theories related to this issue are Piaget's constructivism and Papert's constructionism (Mubin, Stevens, Shahid, Al Mahmud, Dong, 2013). Piaget's constructivism is based on the idea that children, such as adults, do not change their ideas just because an expert tells them they are wrong. Learning is a more complex process that involves interacting with and exploring the world. Knowledge is gained only through an active participation with the world, people and things in a progressive way (Ackermann, 2001). Learners, thus, cannot be considered as a passive component of the learning process. This process is triggered by a state of disequilibrium between a lived experience and our way of thinking. At the beginning, the learner tries to make sense of what he or she is living using his/her current knowledge, but when is no longer able to do it, then the knowledge is restructured bringing it to a higher level of thinking (Amineh, Davatgari Asl, 2015). Papert shares the constructivist view of learning as the construction of evolving meaning structures based on experience. The author also considers the student engagement in the construction of a cognitive artefact as a learning enabler. Papert's constructionism is more situated and pragmatic than Piaget's constructivism (Ackermann, 2001). The learning performance is greater when this process is self-directed and activities are perceived as playful in a way that reproduces how children naturally learn outside of the classroom (Cooper, Martyn; Keating, David; Harwin, Wiliam and Dautenhahn, Kerstin, 1999).

Creating a continuum between the two authors and theories, it is possible to could argue that for both of them **the student is an active component of the learning process** and that **knowledge is not something static to be transmitted, rather something that is personally constructed through the experience** (Ackermann, 2001). Educational Robotics, to lead to a real change, needs **a learning environment based on the above-mentioned educational philosophies** (Alimisis, 2013). Therefore, the learning process moves away from the mere top-down notional transfer, and it is enriched by experience and social interaction (with robots and humans). also allowing the construction and deconstruction of robots as cognitive and social artifacts that permit the development of knowledge in a practical and situated way.

Although their pedagogical thinking does not directly address the application of robotics in educational contexts, authors have taken up the contribution of three pillars such as Piaget, Vigotsky and Bruner to justify the meaning of the landing of robots in the field of human learning and society, more broadly:

- *Piaget*: the importance of the body in relation to technology is critical if technology is a pervasive mediator of the relationship with the world. Sensorimotor developmental stage is based on sensory and bodily experience. Piaget himself, however, recalls how children are great explorers and producers of knowledge: in this sense, technology can be seen as an extraordinary world to offer opportunities for knowledge.
- *Vigotsky*: cultural psychology recontextualizes the digital age discoveries as cultural artifacts. Human artifacts that can direct development on paths that would not be viable by leveraging just our natural conditions. They are expansions of natural possibility.
- *Bruner*: cultural psychology and culture education. Bruner's 3P's for culture education promotion (Past, Present, Possible). In this complex present, it is essential to remember our past because it is an anchor that allows us to live the present time to its fullest (yes opportunities, watch out for risks).

Scope of Application and Types of Robots

In a review conducted back in 2012 by Barreto and Benitti, we can see the first studies related to the introduction of Social Robots in schools, demonstrating that this is still a relatively recent approach, but that at the same time there have already been many experiences. Within the context of their paper, the authors carried out a systematic literature review to examine the state of research in Educational Robotics in schools in the first decade of 2000. Here are two of the research questions that oriented their review (Barreto, Benitti 2010):

- What topics (subjects) are taught through robotics in schools?
- Is robotics an effective tool for teaching? What do the studies show?

The contribution of the authors tends to consider the use and effectiveness of Social Robots in school contexts, allowing us to identify the main topics in which they are used, some types of robots and their main effects on the effectiveness of training / interventions. In a systematic review, Barreto and Benitti (2012) clarify that, among the main subject/knowledge areas identified, mainly **STEM subjects** emerge, related, in particular, to Mathematics (Barker and Ansorge, 2007; Nugent et. al., 2007) for different types of students (Hussain et al, 2006; Lindh & Holgersson, 2007; Hussain et al., 2006) and contents (Nugent et al, 2007), to computing programming (Barker & Ansorge 2007, Nugent et al., 2008), evolution (Whittier and Robinson, 2007), geospatial concept (Barker and Ansorge 2007, Nugent et al., 2008), Physics (Williams et al., 2007; Mitnik et al., 2008), Systems and computation (Sullivan 2008). The authors (Barreto and Benitti, 2012) also outlined a number of **technical and social skills** that are solicited by the use of Social Robots in schools:

- *Thinking skills*: observation, estimation and manipulation (Sullivan, 2008);
- *Science Process skills*: evaluation of solutions; (Sullivan 2008)
- *Hypothesis Generation*; hypothesis testing and control of variables (Sullivan, 2008);
- *Social Interactions* (Mitnik et al., 2008; Owens et al., 2008);
- *Problem Solving Approaches* (Nugent et al., 2009; Hussain et al., 2006; Williams et al., 2007);
- *Skills in conducting Scientific Inquiry* (Williams et al., 2007);
- *Socialization and Communication* (Owens et al., 2008);
- *Teamworking* (Nugent et al., 2009).

Instead, the review conducted by Cheng and Lai (2019) shows its main applications in the field of special education. The authors define Special Education as "an adaptive form of education based on different needs according to ability and is adopted for disabled students with different disability characteristics when they cannot adapt to the regular education system (Florian, 2008)" (Cheng, Lai, 2019, p.2). According to the authors, students with different kind of disabilities have educational needs that differ from those of other students, due to both physical and mental disabilities causing difficulties in learning (Lamsa et al., 2018). In an effort to help this student to reduce the limitations brought by their disabilities, several researchers have specified that technology "support can improve their engagement in learning and their ability to complete learning tasks (Zhang, 2000)" (Cheng, Lai, 2019, p.2). The review (Cheng, Lai, 2019) clearly shows three main potentials of Social Robots, which can be extended to other scenarios as well:

- Social Robots possess characteristics that make them suitable for supporting learning, such as flexibility, interactivity, and humanoid appearance (Marchetti et al., 2018);

- They are fun, activating curiosity and affecting student motivation;
- They foster the development of skills for the 21st century (soft skills, non-technical skills, STEM/STEAM).

These advances in Social Robots resulted in their slow introduction into the educational system (Belpaeme et al., 2018a; Kory-Westlund and Breazeal, 2019) and into homes as well (Scassellati et al., 2018). Indeed, quoting Rosemberg-Kima et. al (2020, p.4) "In recent years, several applications of Social Robots in higher education have started to emerge (Brown and Howard, 2014; Edwards et al., 2016; Deublein et al., 2018)". In a study carried out by Coşkun (2019), the author identified several themes related to the effects of social robots in higher education, specifically concerning awareness about the potential of robots in educational and non-educational areas, development of knowledge (robotic and curricular), approach and increased acceptance towards robots, professional competence and development of efficiency. The use of robots in educational contexts has seen different modalities, related in particular to the type of robot, the educational content and the purpose of their use. It is possible to identify two main categories of robots in educational contexts:

- *Assembly boxes*: (e.g., *Lego Mindstorm*) they allow to build robots and consequently certain educational experiences. Group activities in the construction of the robot develop soft skills such as empathy, designing together, problem solving, negotiation and adaptation within a team. The robots created tend to have typical characteristics of an organism. In addition to L*ego*, there are many other assembly boxes that can be used in soft skills teaching. The use of assembly boxes allows you to act on the architecture, sense-motor system and control system of the robot;
- *Pre-assembled robots*: in this case, the robot is already assembled, so there is no construction phase. There are several types of pre-assembled robots, designed to have appeal for different age groups. Other experiences and possibilities of using robots are elicited here. An example of this type is the NAO robot, or robotic arm. Their use is often directed toward teaching STEM or other curricular subjects through highly engaging experiences.

Regardless of the type of robot used, in both school and training, the areas of use of robots are mainly two:

- *Curricular skills*: using robotics to teach specific STEM-related contents such as, for example, biology (exploiting the robot organism metaphor), mathematics and coding. Here, the two types of robots (assembly boxes and pre-assembled robots) contribute to curricular and notional learning creating engagement and engaging experiences;
- *Soft skills*: in this case the use of the robot aims at developing "soft" skills. The goal is to use the two types of robots to propose group experiences, collaborative and challenging, in which the robot is a vehicle for the relationship.

Robot as a Learning Tool vs. Robot as a Learning Subject

The activities related to robotics in the field of education can be classified into two categories whose discriminant is the role it plays in the learning process: Robotic Education and Educational Robotics. The distinction between Robotic Education and Educational Robotics, although not as clear-cut as it may appear at a first glance, allows to identify two distinct ways of conceiving Social Robots in educational contexts: robots as learning tools and robots as learning objects.

In an approach in which **the robot becomes the primary object of learning**, namely Robotic Education, the goal becomes that of **developing a greater knowledge and familiarity with this specific technology**, thus providing technical and "hard" knowledge. The robot as an object also allows to understand the logic behind intelligent technologies. Today we are in the field of "data" literacy, the frontier has moved towards data. Bringing robotics into education tunes young people into what technologies do and how they work. Finally, it allows supporting the development of digital and cultural competencies that serve to access the world of meaning and to provide the specific knowledge that underpins "Digital Citizenship".

In the second case, considering the **robot as a tool**, corresponding to what authors will call Educational Robotics, **we educate with robotics to promote also other transversal disciplines (Di Dio et al., 2020; Manzi et al., 2020)**. The main advantages consist in the fact that educating with robotics favors an active teaching and all the consequent benefits, it also allows to promote an interactive and collaborative teaching, able to exploit the group and the relationship. The robot as a tool promotes learning based on discovery and problem-solving processes. All of this develops the proximal zone of the human brain. One purpose is also to allow children and adults familiarize with the technological elements of the social system in which they are inserted, increasingly characterized by the relationship between man and machine, as well as by the "sex-appeal" of the inorganic. This familiarization is intended to help children from an early age to develop reflections on what is happening around them and to become "digital citizens". In this second perspective, robotics can be presented as a subject to be studied on its own (e.g., robotics as a learning object), whose teaching involves **carrying out projects allowing to acquire concepts related to Robotics, Engineering and STEM** (Science, Computer Science and Technology) and, at the same time, favoring the **development of transversal skills, namely the 21st century skills**, such as problem solving, creativity, critical thinking, team working (Alimisis, Kynigos, 2009).

Educational Robotics: An Overview

As Rosenberg-Kima, Koren and Gordon (2020, p.2) suggest, "classrooms in the twenty-first century are slowly being transformed from frontal-lecture halls full of passive students to collaborative small groups actively participating in practical learning experiences (Helle et al., 2006; Kokotsaki et al., 2016)". In this sense, **Educational Robotics consists of an engaging learning experience that involves the use of robots for teaching curricular knowledge**, including the construction of a robot itself and/or the programming of already built robots and their control through softwares developed by the students themselves (Denis, Hubert, 2001). It is important to specify that in this vision **the Social Robot is a learning object**, a means to achieve curricular knowledge or skills development goals through innovative, active and engaging processes. The robots used can be assembly boxes as well as pre-assembled Robots.

The great potential associated with Educational Robotics lies in its ability to support the students developing both schools' subjects and what the Fourth Industrial Revolution has brought out as 21st century skills (Alimisis, 2013). These competencies include complex skills such as problem solving, teamworking and collaboration, critical thinking and a high degree of ICT literacy (Yusuf, Walters, Sailin, 2020). The mechanism is so that through the projects of construction, programming and controlling of the robot, students not only acquire technical knowledge related to Technology, Computer Science and Science, but they also confront with each other, reach complex solutions through trial and error processes developing critical thinking and problem solving skills and become more familiar with these robotic devices whose presence will be increasingly pervasive in human life (Toh, Causo, Tzuo, Chen, Yeo, 2016).

When we talk about Robotic Education, we are referring to the use of robots as learning agents that are used to deliver learning experiences through social interactions with different kinds of learners (Belpaeme, Kennedy, Ramachandran, Scassellati, Tanaka, 2018).

Educational robots belong to the wider domain of educational technology and their peculiarity is to improve and facilitate learning by providing embodiment and increase of social interactions in the learning context (Mubin, Stevens, Shahid, Al Mahmud, Dong, 2013). Robots, in fact, thanks to their characteristics of anthropomorphism and movement, are perceived by humans as **social partners** and are therefore able to activate the social areas in our brain (Manzi, Di Dio, Di Lernia, Rossignoli, Maggioni, Massaro, Marchetti, Riva, 2021). However, educational robots are not the only technological tools available to support learning; in fact, one might prefer the use of **virtual agents**, equipped only with the software components and therefore acquired at a lower cost. The choice between these two tools should take into consideration the three main advantages that robots offer over virtual agents:

- *The first advantage* is related to materiality, as robots should be preferred to virtual agents when the topic of study requires physical manipulation of objects or autonomous movement;
- *The second advantage* is related to the key fact that people, interacting with robots, show social behaviors that promote learning, such as greater engagement in activities that require cooperation;
- *The last advantage* is related to the fact that learning through robots affect the way in which students show subsequent behaviors. (Belpaeme, Kennedy, Ramachandran, Scassellati, Tanaka, 2018).

The choice of robots over virtual agents becomes even more useful in the case of students with disabilities: the materiality of robot and its ability to move allow students with physical disabilities to actively participate in works that would otherwise be inaccessible reducing them to mere observers. In this sense, educational robots are a **fundamental tool of inclusion** that makes practical elements of the curriculum accessible to a wide range of students with physical or cognitive limitations (Cooper, Martyn; Keating, David; Harwin, Wiliam and Dautenhahn, Kerstin, 1999).

Belpaeme, Kennedy, Ramachandran, Scassellati, Tanaka (2018) describe **three types of roles** that educational robots play in the learning process:

- *Robot as a Tutor or Teacher Assistant.* The robot has the task of providing suggestions, exercises and supervision. Usually, when the robot has a tutoring role, the learning activity is delivered by means of a one-to-one relationship, in order to maximize the personalization of education; in the case of a teacher robot, or more frequently teacher assistant robot, it is used simply as delivery channel of a lesson shown to attract the attention of the students and to improve their motivation (Belpaeme, Kennedy, Ramachandran, Scassellati, Tanaka, 2018).
- *Robot as a peer.* When the robot plays this role, active spontaneous participation emerges (Mubin, Stevens, Shahid, Al Mahmud, Dong, 2013). In this way, robots present themselves in a less intimidating way than the role of tutor or assistant. Such robots may present themselves as peers with more experience in leading students toward goals that are neither too challenging nor too easy to achieve, or they may configure themselves as robots that offer just motivational support (Belpaeme, Kennedy, Ramachandran, Scassellati, Tanaka, 2018). This focus on the peer role entails a greater emphasis on collaborative, rather than didactic, interactions between the child and the robot (Baxter et al.).
- *Robot as a novice.* Adopting this role in the learning process, it is possible to exploit the so-called *protégé* effect. The student consolidates what he has learned becoming the robot's instructor. This process allows a better understanding and assimilation of the acquired notions, to improve self-confidence and, at the same time, the development of meta-cognitive skills in students due to the fact that they design the teaching process that requires a deeper understanding of what has been studied, but they also deal with the internal representations of robots (Belpaeme, Kennedy, Ramachandran, Scassellati, Tanaka, 2018).

As Koren and Gordon suggest, the choice of the role of educational robots should be based on the learning content, the type of student(s), the teacher, and the nature of the learning activity, ensuring that learning could be as personalized and effective as possible (Mubin, Stevens, Shahid, Al Mahmud, Dong, 2013). This focus has created new various pedagogies, such as to provide students with the opportunity, within the classroom, to observe, imitate and practice critical-agency, and reflect upon it (ten Dam and Volman, 2004);

to collaborate by learning to share tasks and resources being responsible for their tasks (Lai et al., 2017); to be engaged in interdisciplinary approaches promoting creativity (Harris and de Bruin, 2018); to use project-based learning as the basis for improving communication skills (Saenab et al., 2018).

Educational Robotics considers the robot as a tool and not as the object of learning, however it allows a first approach to these technologies in a specific sense and in a broad perspective, starting to shape the so-called Robotic Skills. The case of Comau, that we are going to study in this book, stands as an **Educational Robotics project that revolves around the e.DO Robot**. e.DO is used as a means for learning curricular disciplines through an active and engaging experience, which allows to approach robotics and develop the so-called 21st century skills.

The use of e.Do can also be understood as a form of what Herrington, Reeves, and Oliver (2014) define as Authentic Learning Environments, namely a "pedagogical approach that situates learning tasks in the context of real-world situations" (p 401). Introducing students to professional robotics through e.Do would seem to exemplify this concept.

e.Do is also configured as a form of Problem-based learning (PBL), particularly with regard to the contexts and the projects that students were engaged in. PBL is another pedagogical approach which pairs with Authentic Learning by having students solve an open-ended problem as part of the learning process rather than learning content didactically.

The emotional values that are an outcome of the *e.Do Experience*, are clearly aligned with the Partnership for 21st Century Skills and Social Emotional Learning. According to the Partnership for 21st Century Skills Framework (2008), 21st Century learning skills include collaboration, communication, critical thinking, and creativity.

Social and Emotional Learning is defined as "a process for helping children and even adults develop the fundamental skills for life effectiveness, including, recognizing, and managing emotions, developing caring and concern for others, establishing positive relationships, making responsible decisions, and handling challenging situations constructively and ethically (CASEL, 2013). All of these considerations point to a strong connection between the use of e.Do as a form of SEL and the development of these 21st Century Skills.

Educational Robotics: An Effective Approach

Educational Robotics literature provides several results about the efficacy of this new education approach, even if a more robust empirical support is needed to back it up (Barreto, Benitti, 2012). It is important to underline that, despite the potential of this approach to be applied to different contexts, most of the research findings pertain to the school context; this is probably due to the fact that these new tools and teaching methodologies represent the promise of a contemporary education that allows alignment with the demands of the 4.0 labor market.

Toh, Causo, Tzuo, Chen, Yeo (2016), in their review on the use of robots in education and young children, identify four areas in which Robots and Educational Robotics can positively influence the students' behavior and development:

- *Problem solving abilities, team skills and collaboration*;
- *Achievement scores, science concepts and sequencing skills*;
- *Language skills development*;
- *Participation understood as an increase of students' participation in robotics related activities.*

Barreto and Benitti (2012) point out that, although there is support that Educational Robotics can improve student learning, this is not necessarily the case: there are numerous studies in which the introduction of Educational Robotics had no effect on the students' performance. This leads to the consideration that **Educational Robotics cannot be considered as a *panacea*, it rather should be included and supported by the more general learning context (Barreto, Benitti, 2012).** In this sense, the training and teaching methods of teachers play a crucial role in the success of this educational approach: in fact, it often happens that, due to lack of training, teachers experience discomfort in dealing with such projects because they do not feel prepared to implement them or, sometimes, they struggle to abandon the old teaching model in favor of a new one including a more interactive way of learning and teaching, in accordance with the pedagogical theories of constructivism and constructionism (Baretto, Benitti, 2012; Alimisis, 2019). Research results also support the hypothesis that the learning effects achieved through educational robots lead to higher cognitive and affective outcomes than those of their competitors just equipped with the software component and, furthermore, the use of Social Robots in learning shows positive effects on compliance, engagement and conformity which, in turn, have a positive impact on learning outcomes (Di Dio et al., 2020; Belpaeme, et al., 2018). Educational Robotics, having been used as a tool to prevent school dropout and failure, has also proven to be an **excellent tool for motivation and engagement in learning** (Alimisis, 2019).

In addition, Educational Robotics has been used as a **training activity for pre-service teachers to increase their interest and engagement in STEM** and to improve their method of teaching those subjects (Kim, D. Kim, Yuan, Hill, Doshi, Thai, 2015). The results show that the training developed through Educational Robotics resulted in increased teacher interest and engagement in STEM subjects and robotics as a learning tool and object: participating teachers reported that they understood the benefits of robotics in STEM teaching and were willing to adopt the same approach. All of the teachers involved did not have a high knowledge of robotics and initially they all experienced frustration, but then it turned into enjoyment and then confidence in Educational Robotics (Kim, D. Kim, Yuan, Hill, Doshi, Thai, 2015).

As already mentioned, for the effective introduction of robotics and robots in the educational field, certain contextual conditions must be in place. Among these, one of the most important, especially with regard to the introduction of Social Robots, is the **support from the stakeholders of the process**, especially students, teachers, and parents (Smakman, Jansen, Leunen, Konijn, 2020). Indeed, robots offer many opportunities from an educational perspective but, at the same time, they introduce moral challenges (Smakman, Konij, 2019). The literature needs further investigation of stakeholder perceptions, especially with regard to parents as representatives of children and as those who experience the effects of robot education firsthand (Smakman, Jansen, Leunen, Konijn, 2020). In fact, to truly accept social teaching robots, it is critical that **manufacturing organizations involve and consider these stakeholders in their design processes**. (Smakman, Jansen, Leunen, Konijn, 2020). Smakman and Konijn (2019), through a systematic review of the existing literature, identify 14 moral values that, from teachers and students' perspectives, are affected by the introduction of Social Robots, particularly tutor robots. These are classified into the following categories:

- *Psychological Welfare and Happiness*: robots affect both positively and negatively these two values. For example, robots positively impact these values from the perspective of children by providing them with a funny learning context and, for teachers, robots could be the tool to avoid carrying out boring and repetitive tasks. Conversely, robots could also lead to states of anxiety for children, and teachers show the fear that robots could completely replace them in teaching.
- *Efficiency*: this value could also be positively and negatively affected by the introduction of robots. Although studies show that robots are more effective than their competitors with just the software component, the

problems related to the not yet perfectly fluent interaction could lead to a loss of effectiveness of the learning and teaching processes.

- *Freedom from Bias*: robots, by treating all children equally, are free of social bias, ensuring their absence. Conversely, it is necessary to be alert to the possible presence of technological bias.
- *Usability*: robots can be used by a wide range of users, but there is the problem that not all teachers are able to use them.
- *Deception and Trust*: the robot is perceived as a social partner by the student, who could even develop a friendship relationship with it, but the trust could be damaged by finding out that the robot's data are available to teachers, thus causing the feeling of having been deceived.
- *Friendship and Attachment*: it can be feared that the relationship established with the robot may undermine the concepts of friendship and attachment of the child.
- *Human Contact*: it is feared that children may prefer the company of a robot rather than of a human, thus reducing human contact.
- *Privacy, Security, Safety and Accountability*: this refers mainly to the uncertainty and fear associated with the management of data recorded by the robot.

Smakman, Jansen, Leunen, and Konijn (2020), broadening the scope of the analysis, extend this study by incorporating the parental perspective. From the parents' perspective, robots influence nine of the above values: Psychological welfare and happiness, Efficiency, Usability, Human contact and Privacy, protection, security and accountability (Smakman, Jansen, Leunen, Konijn, 2020). Overall, parents see as opportunities the patient behavior of robots, their ability to promote kindness, their potential to assist their children's learning even at home and their ability to enhance the children's learning. The most important concerns are here again related to privacy, security, safety and accountability issues, with a focus on data management. However, three additional concerns emerged from this analysis: the first, inherent to the moral value of flexibility, is the parents' fear that robots will become a baggage to be carried everywhere; the second and third ones are instead related to the values of responsibility and dependability, as parents suspect a shift in responsibility and excessive dependence on robots, as well as that the robots will take on a parental role.

Generally speaking, it can be said that **all stakeholders, although they have concerns, see great opportunities in the use of robots in the school context**: children perceive them as playful and engaging partners, teachers understand their potential for teaching and parents correlate them with their children's higher learning performance (Toh, Causo, Tzuo, Chen, Yeo, 2016). It would be interesting, however, to continue investigating these kinds of perceptions, especially with regard to the parents' ones, rarely studied (Smakman, Jansen, Leunen, Konijn, 2020).

CONCLUSION

As Rosenberg-Kima, Koren and Gordon (2020, p.2) suggest, "classrooms in the twenty-first century are slowly being transformed from frontal-lecture halls full of passive students to collaborative small groups actively participating in practical learning experiences (Helle et al., 2006; Kokotsaki et al., 2016)". Studies have shown that such active participatory learning is more effective in content retention (Al-Balushi and Al-Aamri, 2014) and engagement (Fernandes et al., 2014). Thus, emphasis has been diverted to so-called 21st century skills (Crane, 2003; Saavedra and Opfer, 2012; Trilling and Fadel, 2012), with focus on Communication, Collaboration, Creativity, and Critical thinking (Shulman, 1986; Kivunja, 2015).

Robots as engagement facilitator in classroom activities has therefore potentially **significant consequences for social developement, as well as for learning (Rosenberg-kima et al. 2020)**. Rosembertg-Kima et al. (2020) point out that the presence of social robots brings several advantages over other media for behavioral change and learnings in students. Different studies highlight that physical robots will attract more attention, and will comply with their requests; following evidence suggest children consider Robots as psychological and social agents, perceived as more enjoyable interaction partners **(Rosenberg-kima et al.2020)**.

Here are many recent studies showing that **the use of social robots in education supports learning from different point of view**:

- The effect of some **psychological variables on learning**: social robots have the potential to increase the students' computational thinking skills (Sullivan & Bers, 2015), self-study skills (Goh & Aris, 2007), teamworking skills (Nourbakhsh et al., 2005), problem-solving skills (Robinson, 2005; Rogers & Portsmore, 2004), collaborative learning and self-confidence, communication, and critical thinking skills (Saygin, Yuen, Shipley, Wan, & Akopian, 2012).

- Social robot as **a supportive learning tool for teachers**: Social Robots can support educational environments by triggering curiosity (Hidi & Renninger, 2006), increasing engagement (Miller & Stein, 2000; Nourbakhsh et al., 2005), enabling real-world experiences (Nalajala, 2003), basic robotics and programming skills (Sullivan & Bers, 2015).
- Social Robots have also been defined as **an effective tool to be used in the fields of STEM** (Johnson, 2003). The study of Sullivan and Bers (2015) has point out that "using robotics and computer programming in early childhood education can support the development of a range of cognitive and social milestones" (**Rosenberg-kima et al.2020, p.7**). Robots introduce students to **new and innovative technologies**, and they also encourage them to become **active creators** of technological products in the future (Eguchi, 2014).

So, at the end of the desk-based research related to Educational Robotics, authors have reorganized its main implications based on **4 areas of value generated for those who take part in it**; these are presented below.

1. Skill Creation

Educational robotics lends itself as an extremely effective way to develop **enabling skills for living and working in the 4.0 world**, such are **key skills in the Digital Transformation scenario** we live in, as emerged in the previous chapter. Specifically, authors have identified a number of skills that are transversally found in this type of initiatives. They can be summarized as follows:

- *Robotic Skills*; skills related to the programming and use of robots, the ability to relate to them and to grasp their potential.
- *School Subject Skills*; development of specific school and curriculum knowledge primarily related to STEM Skills.
- *Working in the Future Skills*; enablement of skills to work in the world of work of the future, covering the ability to relate to other individuals and new technologies, agility, engagement, collaboration, interdisciplinarity and innovation.
- *Professional Skills*; development of technical and operational skills related to new technologies to facilitate the entry into the world of work.

2. Value for New Citizenship

Reference is made here to the skills needed to **develop a sense of responsible and just citizenship**. In this area authors find the dimension of challenge elicited in the initiatives of Educational Robotics, the stimulus to leave one's comfort zone during contact with new technologies and unusual ways of learning, the drive to respect others, to keep promises, to be exemplary and consistent, as well as to measure results objectively.

3. Empowerment

Educational Robotics initiatives seem to be proactive in creating value in terms of "empowerment", as they become a means of **becoming aware of one's potential, strengths and areas for improvement**. They also seem to be particularly suitable for the development of new skills, knowledge, and abilities, as well as to stimulate the proposal of new solutions and initiatives to respond to technical problems and / or daily life.

4. Inspiration and Orienteering for Working

The initiatives of Educational Robotics also serve to show a new view of the social reality and the world of work, characterized by the 4.0 environment and the changes in Digital Transformation. In this sense, they offer ideas to **expand knowledge and awareness concerning the world of work of the future**, increasingly populated by new technologies. They are also a chance to discover new passions or areas of interest in this field and to **identify possible career paths and/or study** related to Digital Transformation.

The next chapters will take into consideration **a real Educational Robotics project**, the Comau *e.DO Experience*, to verify and explore the areas of value identified through the desk-based research activity.

REFERENCES

Ackermann, E. (2001). *Piaget's constructivism, Papert's constructionism: What's the difference.* Future of Learning Group Publication.

Al-Balushi, S. M., & Al-Aamri, S. S. (2014). The effect of environmental science projects on students' environmental knowledge and science attitudes. *International Research in Geographical and Environmental Education*, *23*(3), 213–227. doi:10.1080/10382046.2014.927167

Alimisis, D. (2013). Educational robotics: Open questions and new challenges. *Themes in Science and Technology Education*, *6*(1), 63–71.

Alimisis, D. (2019). Teacher training in educational robotics: The ROBOESL Project paradigm. *Technology. Knowledge and Learning*, *24*(2), 279–290. doi:10.100710758-018-9357-0

Alimisis, D., Alimisi, R., Loukatos, D., & Zoulias, E. (2019). Introducing maker movement in educational robotics: beyond prefabricated robots and "black boxes". In L. Daniela (Ed.), *Smart Learning with Educational Robotics, Using Robots to Scaffold Learning Outcomes* (pp. 93–115). Springer. doi:10.1007/978-3-030-19913-5_4

Alves-Oliveira, P., Arriaga, P., Paiva, A., & Hoffman, G. (2019). Guide to build YOLO, a creativity-stimulating robot for children. *HardwareX*, *6*, e00074. doi:10.1016/j.ohx.2019.e00074

Alves-Oliveira, P., Sequeira, P., Melo, F. S., Castellano, G., & Paiva, A. (2019). Empathic robot for group learning: A field study. *ACM Transactions on Human-Robot Interaction*, *8*(1), 1–34. doi:10.1145/3300188

Amineh, R. J., & Asl, H. D. (2015). Review of constructivism and social constructivism. *Journal of Social Sciences. Literature and Languages*, *1*(1), 9–16.

Arís, N., & Orcos, L. (2019). Educational robotics in the stage of secondary education: Empirical study on motivation and STEM skills. *Education Sciences*, *9*(2), 73. doi:10.3390/educsci9020073

Baraka, K., Alves-Oliveira, P., & Ribeiro, T. (2020). An extended framework for characterizing social robots. In *Human-Robot Interaction* (pp. 21–64). Springer. doi:10.1007/978-3-030-42307-0_2

Barker, B. S., & Ansorge, J. (2007). Robotics as means to increase achievement scores in an informal learning environment. *Journal of Research on Technology in Education*, *39*(3), 229–243. doi:10.1080/15391523.2007.10782481

Baxter, R., & Richard, B. (1994). *Baxter: A Holy Commonwealth*. Cambridge University Press. doi:10.1017/CBO9781139170260

Belpaeme, T., Kennedy, J., Baxter, P., Vogt, P., Krahmer, E. E., Kopp, S., Bergmann, K., Leseman, P., Küntay, A. C., Göksun, T., Pandey, A. K., Gelin, R., & Deblieck, T. (2015). L2TOR-second language tutoring using social robots. *Proceedings of the ICSR 2015 WONDER Workshop*.

Belpaeme, T., Kennedy, J., Ramachandran, A., Scassellati, B., & Tanaka, F. (2018). Social robots for education: A review. *Science Robotics*, *3*(21).

Benitti, F. B. V. (2012). Exploring the educational potential of robotics in schools: A systematic review. *Computers & Education*, *58*(3), 978–988.

Billard, A., & Dautenhahn, K. (1997). Grounding communication in situated, social robots. In *Proceedings Towards Intelligent Mobile Robots Conference*. Report No. UMCS-97-9-1, Department of Computer Science, Manchester University.

Björling, E. A., & Rose, E. (2019). Participatory research principles in human-centered design: Engaging teens in the co-design of a social robot. *Multimodal Technologies and Interaction*, *3*(1), 8.

Brown, L. N., & Howard, A. M. (2014, March). The positive effects of verbal encouragement in mathematics education using a social robot. In *2014 IEEE integrated STEM education conference* (pp. 1-5). IEEE

Calderita, L. V., Manso, L. J., Bustos, P., Suárez-Mejías, C., Fernández, F., & Bandera, A. (2014). THERAPIST: towards an autonomous socially interactive robot for motor and neurorehabilitation therapies for children. *JMIR Rehabilitation and Assistive Technologies, 1*(1), e3151.

Ceha, J., Chhibber, N., Goh, J., McDonald, C., Oudeyer, P. Y., Kulić, D., & Law, E. (2019, May). Expression of curiosity in social robots: design, perception, and effects on behaviour. In *Proceedings of the 2019 CHI Conference on Human Factors in Computing Systems* (pp. 1-12). ACM.

Collaborative for Academic, Social, and Emotional Learning (CASEL). (2013). *Success in School. Skills for life*. Retrieved from www.casel.org

Cooper, M., Keating, D., Harwin, W., & Dautenhahn, K. (1999). Robots in the classroom-tools for accessible education. Assistive Technology on the Threshold of the New Millennium, 448–452.

Correia, F., Petisca, S., Alves-Oliveira, P., Ribeiro, T., Melo, F. S., & Paiva, A. (2019). "I Choose... YOU!" Membership preferences in human–robot teams. *Autonomous Robots*, *43*(2), 359–373.

Correia, N., Camilo, C., Aguiar, C., & Amaro, F. (2019). Children's right to participate in early childhood education settings: A systematic review. *Children and Youth Services Review*, *100*, 76–88.

Coşkun, S., Kayıkcı, Y., & Gençay, E. (2019). Adapting engineering education to industry 4.0 vision. *Technologies*, *7*(1), 10.

Crane, A., Matten, D., & Moon, J. (2003, July). Can Corporations be Citizens?: Corporate citizenship as a metaphor for business participation in society. In *Proceedings of the International Association for Business and Society* (*Vol. 14*, pp. 105-115). Academic Press.

Denis, B., & Hubert, S. (2001). Collaborative learning in an educational robotics environment. *Computers in Human Behavior*, *17*(5-6), 465–480.

Deublein, A., Pfeifer, A., Merbach, K., Bruckner, K., Mengelkamp, C., & Lugrin, B. (2018). Scaffolding of motivation in learning using a social robot. *Computers & Education*, *100*(125), 182–190.

Di Dio, C., Manzi, F., Peretti, G., Cangelosi, A., Harris, P. L., Massaro, D., & Marchetti, A. (2020). Come i bambini pensano alla mente del robot. Il ruolo dell'attaccamento e della Teoria della Mente nell'attribuzione di stati mentali ad un agente robotico. *Sistemi Intelligenti*, *32*(1), 41–56.

Edwards, A., Edwards, C., Spence, P. R., Harris, C., & Gambino, A. (2016). Robots in the classroom: Differences in students' perceptions of credibility and learning between "teacher as robot" and "robot as teacher". *Computers in Human Behavior*, *65*, 627–634.

Eguchi, A. (2014, July). Robotics as a learning tool for educational transformation. In *Proceeding of 4th international workshop teaching robotics, teaching with robotics & 5th international conference robotics in education Padova (Italy)* (pp. 27-34). Academic Press.

Florian, L. (2008). Inclusion: special or inclusive education: future trends. *British Journal of Special Education*, *35*(4), 202–208.

Giger, J. C., Piçarra, N., Alves-Oliveira, P., Oliveira, R., & Arriaga, P. (2019). Humanization of robots: Is it really such a good idea? *Human Behavior and Emerging Technologies*, *1*(2), 111–123.

Goh, H., & Aris, B. (2007, September). Using robotics in education: Lessons learned and learning experiences. In *1st International Malaysian Educational Technology Convention* (pp. 1156-1163). Academic Press.

Gordon, G., Spaulding, S., Westlund, J. K., Lee, J. J., Plummer, L., Martinez, M., Das, M., & Breazeal, C. (2016, March). Affective personalization of a social robot tutor for children's second language skills. *Proceedings of the AAAI Conference on Artificial Intelligence*, *30*(1).

Goudzwaard, M., Smakman, M., & Konijn, E. A. (2019, August). Robots are good for profit: A business perspective on robots in education. In *2019 Joint IEEE 9th International Conference on Development and Learning and Epigenetic Robotics (ICDL-EpiRob)* (pp. 54-60). IEEE.

Guerin, B., Janta, B., van Gorp, A. (2018). Desk-based research and literature review. *Evaluating interventions that prevent or counter violent extremism*, 63.

Harris, A., & De Bruin, L. R. (2018). Secondary school creativity, teacher practice and STEAM education: An international study. *Journal of Educational Change*, *19*(2), 153–179.

Hein, M., & Nathan-Roberts, D. (2018, September). Socially interactive robots can teach young students language skills; a systematic review. *Proceedings of the Human Factors and Ergonomics Society Annual Meeting*, *62*(1), 1083–1087.

Helle, L., Tynjälä, P., & Olkinuora, E. (2006). Project-based learning in post-secondary education–theory, practice and rubber sling shots. *Higher Education*, *51*(2), 287–314.

Herrington, J., Reeves, T. C., & Oliver, R. (2014). Authentic learning environments. In *Handbook of research on educational communications and technology* (pp. 401–412). Springer.

Hidi, S., & Renninger, K. A. (2006). The four-phase model of interest development. *Educational Psychologist*, *41*(2), 111–127.

Hussain, S., Xie, S. Q., & Liu, G. (2011). Robot assisted treadmill training: Mechanisms and training strategies. *Medical Engineering & Physics*, *33*(5), 527–533.

Johal, W., Castellano, G., Tanaka, F., & Okita, S. (2018). Robots for learning.

Johnson, C. C. (2003). *Barriers influencing implementation of the National Science Education Standards by middle school teachers engaged in collaborative professional development*. University of Cincinnati.

Jung, M. F., DiFranzo, D., Stoll, B., Shen, S., Lawrence, A., & Claure, H. (2018). *Robot assisted tower construction-a resource distribution task to study human-robot collaboration and interaction with groups of people*. arXiv preprint arXiv:1812.09548.

Jung, M. F., Martelaro, N., & Hinds, P. J. (2015). Using robots to moderate team conflict: The case of repairing violations. In *Proceedings of the tenth annual ACM/IEEE international conference on Human-Robot Interaction* (pp. 229–236). ACM.

Kane, G. C. (2015). Enterprise social media: Current capabilities and future possibilities. *MIS Quarterly Executive*, *14*(1).

Kim, C., Kim, D., Yuan, J., Hill, R. B., Doshi, P., & Thai, C. N. (2015). Robotics to promote elementary education pre-service teachers' STEM engagement, learning, and teaching. *Computers & Education*, *91*, 14–31.

Kim, E. S., Berkovits, L. D., Bernier, E. P., Leyzberg, D., Shic, F., Paul, R., & Scassellati, B. (2013). Social robots as embedded reinforcers of social behavior in children with autism. *Journal of Autism and Developmental Disorders*, *43*(5), 1038–1049.

Kivunja, C. (2015). Teaching students to learn and to work well with 21st century skills: Unpacking the career and life skills domain of the new learning paradigm. *International Journal of Higher Education*, *4*(1), 1–11.

Kokotsaki, D., Menzies, V., & Wiggins, A. (2016). Project-based learning: A review of the literature. *Improving Schools*, *19*(3), 267–277.

Konijn, E. A., Smakman, M., & van den Berghe, R. (2020). Use of robots in education. The International Encyclopedia of Media Psychology, 1-8.

Kory, J., & Breazeal, C. (2014). Storytelling with robots: Learning companions for preschool children's language development. In *The 23rd IEEE international symposium on robot and human interactive communication* (pp. 643-648). IEEE.

Kory-Westlund, J. M., & Breazeal, C. (2019). A long-term study of young children's rapport, social emulation, and language learning with a peer-like robot playmate in preschool. *Frontiers in Robotics and AI*, *6*, 81.

Kucuk, S., & Sisman, B. (2018). Pre-Service Teachers' Experiences in Learning Robotics Design and Programming. *Informatics in Education, 17*(2), 301–320.

Kucuk, S., & Sisman, B. (2020). Students' attitudes towards robotics and STEM: Differences based on gender and robotics experience. *International Journal of Child-Computer Interaction, 23*, 100167.

Lai, R., Lin, W., & Wu, Y. (2018, August). Review of research on the key technologies, application fields and development trends of intelligent robots. In *International conference on intelligent robotics and applications* (pp. 449-458). Springer.

Lämsä, J., Hämäläinen, R., Aro, M., Koskimaa, R., & Äyrämö, S. M. (2018). Games for enhancing basic reading and maths skills: A systematic review of educational game design in supporting learning by people with learning disabilities. *British Journal of Educational Technology, 49*(4), 596–607.

Lave, J., & Wenger, E. (1990). *Situated learning: Legitimate peripheral participation.* Cambridge University Press.

Leite, E., Correia, E. B., Sánchez-Fernández, M. D., & Leite, E. (2015). El espíritu emprendedor: Condicionantes para la innovación. *Holos, 5*, 278–291.

Leyzberg, D., Spaulding, S., Toneva, M., & Scassellati, B. (2012). The physical presence of a robot tutor increases cognitive learning gains. In *Proceedings of the annual meeting of the cognitive science society* (*Vol. 34*). Academic Press.

Li, Y., Sam Ge, S., & Yang, C. (2012). Learning impedance control for physical robot–environment interaction. *International Journal of Control, 85*(2), 182–193.

Lindh, J., & Holgersson, T. (2007). Does lego training stimulate pupils' ability to solve logical problems? *Computers & Education, 49*(4), 1097–1111.

Luria, M., Hoffman, G., & Zuckerman, O. (2017, May). Comparing social robot, screen and voice interfaces for smart-home control. In *Proceedings of the 2017 CHI conference on human factors in computing systems* (pp. 580-628). ACM.

Manzi, F., Di Dio, C., Di Lernia, D., Rossignoli, D., Maggioni, M. A., Massaro, D., Marchetti, A., & Riva, G. (2021). Can you activate me? from robots to human brain. *Frontiers in Robotics and AI, 8*, 14.

Manzi, F., Peretti, G., Di Dio, C., Cangelosi, A., Itakura, S., Kanda, T., Ishiguro, H., Massaro, D., & Marchetti, A. (2020). A robot is not worth another: Exploring children's mental state attribution to different humanoid robots. *Frontiers in Psychology*, 11.

Marchetti, A., Manzi, F., Itakura, S., & Massaro, D. (2018). Theory of mind and humanoid robots from a lifespan perspective. *Zeitschrift fur Psychologie mit Zeitschrift fur Angewandte Psychologie*.

Matsuyama, Y., Akiba, I., Fujie, S., & Kobayashi, T. (2015). Four-participant group conversation: A facilitation robot controlling engagement density as the fourth participant. *Computer Speech & Language*, *33*(1), 1–24.

Matsuyama, Y., Taniyama, H., Fujie, S., & Kobayashi, T. (2008). Designing communication activation system in group communication. In *Humanoids 2008-8th IEEE-RAS International Conference on Humanoid Robots* (pp. 629-634). IEEE.

Miller, D. P., & Stein, C. (2000). "So that's what Pi is for!" and other educational epiphanies from hands-on robotics. In Robots for kids: Exploring new technologies for learning (pp. 219-243). Academic Press.

Mitnik, R., Nussbaum, M., & Soto, A. (2008). An autonomous educational mobile robot mediator. *Autonomous Robots*, *25*(4), 367–382.

Mubin, O., Stevens, C. J., Shahid, S., Al Mahmud, A., & Dong, J. J. (2013). A review of the applicability of robots in education. *Journal of Technology in Education and Learning, 1*(209), 13.

Nalajala, T. D. (2003). *Internet telerobotic system*. University of Nevada.

Nourbakhsh, I. R., Crowley, K., Bhave, A., Hamner, E., Hsiu, T., Perez-Bergquist, A., Richards, S., & Wilkinson, K. (2005a). The robotic autonomy mobile robotics course: Robot design, curriculum design and educational assessment. *Autonomous Robots*, *18*(1), 103–127.

Nourbakhsh, I. R., Sycara, K., Koes, M., Yong, M., Lewis, M., & Burion, S. (2005). Human-robot teaming for search and rescue. *IEEE Pervasive Computing*, *4*(1), 72–79.

Nugent, C., Mulvenna, M., Moelaert, F., Bergvall-Kåreborn, B., Meiland, F., Craig, D., & Bengtsson, J. E. (2007, June). Home based assistive technologies for people with mild dementia. In *International Conference on Smart Homes and Health Telematics* (pp. 63-69). Springer.

Nugent, C. D., Finlay, D. D., Fiorini, P., Tsumaki, Y., & Prassler, E. (2008). Editorial home automation as a means of independent living. *IEEE Transactions on Automation Science and Engineering*, *5*(1), 1–9.

Okita, S. Y., Ng-Thow-Hing, V., & Sarvadevabhatla, R. (2009, September). Learning together: ASIMO developing an interactive learning partnership with children. In *RO-MAN 2009-The 18th IEEE International Symposium on Robot and Human Interactive Communication* (pp. 1125-1130). IEEE.

Owens, G., Granader, Y., Humphrey, A., & Baron-Cohen, S. (2008). LEGO® therapy and the social use of language programme: An evaluation of two social skills interventions for children with high functioning autism and Asperger syndrome. *Journal of Autism and Developmental Disorders*, *38*(10), 1944–1957.

Park, H. W., Grover, I., Spaulding, S., Gomez, L., & Breazeal, C. (2019, July). A model-free affective reinforcement learning approach to personalization of an autonomous social robot companion for early literacy education. *Proceedings of the AAAI Conference on Artificial Intelligence*, *33*(01), 687–694.

Park, H. W., Rosenberg-Kima, R., Rosenberg, M., Gordon, G., & Breazeal, C. (2017, March). Growing growth mindset with a social robot peer. In *Proceedings of the 2017 ACM/IEEE international conference on human-robot interaction* (pp. 137-145). ACM.

Partnership for 21st Century Skills. (2008). *A Report and Mile Guide for 21st Century Skills*. Retrieved September 10, 2008 from: http://www.battelleforkids.org/networks/p21

Rogers, C., & Portsmore, M. (2004). Bringing engineering to elementary school. *Journal of STEM Education: Innovations and Research*, *5*(3).

Rosenberg-Kima, R. B., Koren, Y., & Gordon, G. (2020). Robot-supported collaborative learning (rscl): Social robots as teaching assistants for higher education small group facilitation. *Frontiers in Robotics and AI*, *6*, 148.

Saavedra, A. R., & Opfer, V. D. (2012). Learning 21st-century skills requires 21st-century teaching. *Phi Delta Kappan*, *94*(2), 8–13.

Saenab, S., Yunus, S. R., Saleh, A. R., Virninda, A. N., Hamka, L., & Sofyan, N. A. (2018, June). Project-based learning as the atmoshphere for promoting students' communication skills. *Journal of Physics: Conference Series*, *1028*(1), 012026.

Sapounidis, T., & Alimisis, D. (2020). *Educational robotics for STEM: a review of technologies and some educational considerations. In Science and Mathematics Education for 21st Century Citizens: Challenges and Ways Forward.* Nova Science Publishers.

Saygin, C., Yuen, T. T., Shipley, H. J., & Akopian, D. (2012, June). Design, development, and implementation of educational robotics activities for k-12 students. In *2012 ASEE Annual Conference & Exposition* (pp. 25-404). Academic Press.

Scassellati, B., Boccanfuso, L., Huang, C. M., Mademtzi, M., Qin, M., Salomons, N., Ventola, P., & Shic, F. (2018). Improving social skills in children with ASD using a long-term, in-home social robot. *Science Robotics*, *3*(21).

Scassellati, B., Brawer, J., Tsui, K., Nasihati Gilani, S., Malzkuhn, M., Manini, B., Stone, A., Kartheiser, G., Merla, A., Shapiro, A., Traum, D., & Petitto, L. A. (2018, April). Teaching language to deaf infants with a robot and a virtual human. In *Proceedings of the 2018 CHI Conference on human Factors in computing systems* (pp. 1-13). ACM.

Shiomi, M., Kanda, T., Howley, I., Hayashi, K., & Hagita, N. (2015). Can a social robot stimulate science curiosity in classrooms? *International Journal of Social Robotics*, *7*(5), 641–652.

Shulman, L. S. (1986). Those who understand: Knowledge growth in teaching. *Educational Researcher*, *15*(2), 4–14.

Sisman, B., Kucuk, S., & Yaman, Y. (2021). The effects of robotics training on children's spatial ability and attitude toward STEM. *International Journal of Social Robotics*, *13*(2), 379–389.

Smakman, M., Jansen, B., Leunen, J., & Konijn, E. A. (2020, March). Acceptable Social Robots in Education: A Value Sensitive Parent Perspective. In *Proceedings of the INTED2020 Conference* (pp. 7946-7953). Academic Press.

Smakman, M., & Konijn, E. A. (2019, April). Robot tutors: Welcome or ethically questionable? In *International Conference on Robotics in Education (RiE)* (pp. 376-386). Springer.

Spaulding, S., Gordon, G., & Breazeal, C. (2016, May). Affect-aware student models for robot tutors. In *Proceedings of the 2016 International Conference on Autonomous Agents & Multiagent Systems* (pp. 864-872). Academic Press.

Strohkorb, S., Leite, I., Warren, N., & Scassellati, B. (2015, November). Classification of children's social dominance in group interactions with robots. In *Proceedings of the 2015 ACM on International Conference on Multimodal Interaction* (pp. 227-234). ACM.

Sullivan, A., Elkin, M., & Bers, M. U. (2015, June). KIBO robot demo: engaging young children in programming and engineering. In *Proceedings of the 14th international conference on interaction design and children* (pp. 418-421). Academic Press.

Sullivan, F. R. (2008). Robotics and science literacy: Thinking skills, science process skills and systems understanding. *Journal of Research in Science Teaching: The Official Journal of the National Association for Research in Science Teaching, 45*(3), 373–394.

Sundqvist, A., Koch, F. S., Birberg Thornberg, U., Barr, R., & Heimann, M. (2021). Growing Up in a Digital World–Digital Media and the Association With the Child's Language Development at Two Years of Age. *Frontiers in Psychology, 12*, 443.

Tahir, Y., Rasheed, U., Dauwels, S., & Dauwels, J. (2014). Perception of humanoid social mediator in two-person dialogs. In *Proceedings of the 2014 ACM/IEEE International Conference on Human-robot Interaction*. ACM.

Ten Dam, G., & Volman, M. (2004). Critical thinking as a citizenship competence: Teaching strategies. *Learning and Instruction, 14*(4), 359–379.

Toh, L. P. E., Causo, A., Tzuo, P. W., Chen, I. M., & Yeo, S. H. (2016). A review on the use of robots in education and young children. *Journal of Educational Technology & Society, 19*(2), 148–163.

Trilling, B., & Fadel, C. (2012). Tactics for success. *RSA Journal, 158*(5550), 10–15.

Vogt, P., van den Berghe, R., de Haas, M., Hoffman, L., Kanero, J., Mamus, E., Montainer, J.-M., Oranç, C., Oudgenoeg-Paz, O., Garcia., D., H, Papadopulos., F., Schodde, T., Verhagen, J., Wallbridgell, C, D., Willemsen, B., de Wit, J., Belpaeme, T., Göksun, T., Kopp, S., Krahmer, E., Küntay, A, C., Leseman, P., & Pandey, A. K. (2019, March). Second language tutoring using social robots: a large-scale study. In *2019 14th ACM/IEEE International Conference on Human-Robot Interaction (HRI)* (pp. 497-505). IEEE.

Wainer, J., Feil-Seifer, D. J., Shell, D. A., & Mataric, M. J. (2006, September). The role of physical embodiment in human-robot interaction. In *ROMAN 2006-The 15th IEEE International Symposium on Robot and Human Interactive Communication* (pp. 117-122). IEEE.

Whittier, L. E., & Robinson, M. (2007). Teaching evolution to non-English proficient students by using Lego robotics. *American Secondary Education*, 19–28.

Yusuf, B., Walters, L. M., & Sailin, S. N. (2020). Restructuring Educational Institutions for Growth in the Fourth Industrial Revolution (4IR): A Systematic Review. *International Journal of Emerging*.

Chapter 4
e.DO Experience Project

ABSTRACT

In this chapter, the authors present the e.DO Experience Project. This project has many elements of interest with respect to the scope of the research described in the previous pages, and it has been chosen to verify what emerged from desk-based research, with particular reference to the four areas of value identified. In fact, educational robotics initiatives seem to serve several functions, particularly related to their effects on the participants: skill creation, value for new citizenship, empowerment, and inspiration and orienteering for work. The opportunity to analyze in depth an educational robotics project promoted by a reality immersed in the Fourth Industrial Revolution such as Comau through its educational robot e.DO—which is aimed at an extremely transversal population, from kids to professionals—allowed the authors to verify the presence of these areas of value and to investigate any other levels of value generated by the project itself.

INTRODUCTION

Comau is a leading company in the industrial automation field, at a global level. Combining innovative engineering solutions with easy to use, open automation and enabling technologies, it helps companies of all sizes - and across a wide range of industrial segments - leveraging the full potential of digital manufacturing.

Comau's competency stems from over 45 years of field proven-experience and a strong presence within every major industrial country. The company

DOI: 10.4018/978-1-7998-8653-2.ch004

has its headquarters in Turin and operates through an international network of 7 innovation centers, 5 digital hubs, 8 production plants, employing over 9,000 people in 14 countries. Comau develops Industry 4.0-enabled systems, products and services. The full portfolio includes joining, assembly and machining solutions for traditional and electric vehicles, robotized manufacturing systems, a complete family of robots (including collaborative and wearable robotics) with extensive range and payload configurations, autonomous logistics and asset optimization services with real-time monitoring and control capabilities. The offering also extends to project management and consultancy, IoT services, as well as to maintenance and training for a wide range of industrial segments. Innovation is a recurring theme that accompanies almost every activity within Comau - from products to systems and from training to customer relations - and extends to spearheading a new era of automation that is open, connected and easy to understand and to use. Aided by the winning combination of **advanced technological know-how** and **long-term experience**, Comau innovation is driving the development of new products, cutting-edge technologies and a powerful concept of the factory that optimizes **human and machine-driven automation**, called HUMANufacturing (Salvadore, 2019). According to this concept, the factory of the future must be characterized by:

- strong collaboration between man and robot;
- industrial automation accessible to the end user;
- flexibility, use of advanced materials, achievement of high performance in terms of quality and productivity, sustainable production.

Comau has always considered its role to be central in terms of **social responsibility towards the stakeholders** who are part of its ecosystem. By embracing an approach that points to the creation of shared value in a generative and community-oriented perspective (Staerklé et al., 2015), Comau aims to contribute to the **dissemination of technological knowledge and to a business culture capable of leading the transformation of the economic and social system within the Fourth Industrial Revolution**. Through its Academy - conceived as a **real learning factory** - Comau wants to share its knowledge by offering effective solutions to improve skills and business opportunities for people and organizations by providing them with masters, training programs, seminars and publications.

This chapter proposes therefore to describe the characteristics of the **Comau Academy** and to analyze one of its most interesting Educational Robotics

Project - ***e.DO Experience*** - to focus on its specificities. Subsequently, authors will verify the presence of the areas of values generated by this project to understand if they are the same as those identified thanks to the previous desk-based research.

THE COMAU ACADEMY

The Comau Academy was founded in 2007 to accompany the development of the company's internal skills and since 2012 it has been opening to the outside world (Figure 1). The Academy was originally conceived as **a business support tool**, and it firstly launched educational activities in the field of Project Management to develop the skills of its workforce. In the subsequent years, it has gradually extended its initiatives, driven by the logic of **social responsibility, branding and business development**, in line with a generative and common based CVS approach (Porter, Kramer 2011).

Figure 1. Comau Academy evolution

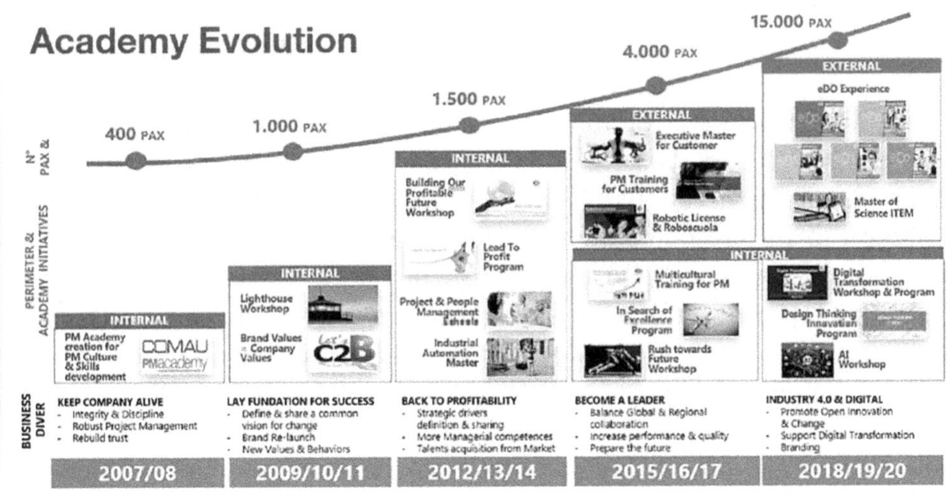

Between 2012 and 2014, the Comau Academy created the *Project & People Management School* and the *Industrial Automation* Master, for the selection and training of talents from the external market. In 2015 the Academy created the *Manufacturing Automation & Digital Transformation* Executive Master with ESCP Business School, which allows executives and managers to explore the issues of manufacturing automation and Digital Transformation. In the same period, the Comau Academy also began to address schools, creating the *Robotics Patent* (*Patentino della Robotica)* project together with Pearson Publishing House, a certification for the use and programming of industrial robots with international significance. The initiatives follow one another and thanks to them the Comau Academy gives life to **a real training ecosystem** (Pinto et al., 2021), within which it designs and implements increasingly **specific and innovative educational paths** in collaboration with business schools, universities, research centers, training institutions and schools of excellence, both nationally and internationally.

The activities of the Comau Academy extend to the dissemination of skills on topics such as **digital transformation**, **technological and business innovation**, **man-machine collaboration**, according to a "Human & Technology based" approach. Highly specialized programs are also launched, such as the *Manufacturing 4.0* Master and the *Master of Science in Innovation & Technology Management*, intended for university students who wish to deepen knowledge related to innovative methodologies, business processes and enabling technologies - from the Industrial Internet of Things (IoT) to collaborative robotics, up to Additive and Digital Manufacturing. Today, the Academy is a **leader in the development of industrial skills**: together with some of the most renowned educational institutions in the world, it offers a wide range of training programs, including Executive Masters and Specialized Masters, Degree and Postgraduate Masters, Qualifications and Professional Certifications. With its programs, the Comau Academy has the ambition to inspire companies, professionals, young talents, schools, and universities. From 2012 to today, the Comau Academy has trained more than 20,000 people of all ages and backgrounds, involving over 320 internal and external teachers in its educational paths. To achieve these objectives, it has collaborated with more than 600 different schools, universities and business schools, involving in its programs a faculty made up of more than 500 teachers and SMEs (Subject Matter Experts) belonging to its network.

All the initiatives designed by the Comau Academy are based on three fundamental elements: the **Comau Change Model**, the **Comau Learning Model** and the **Partnership Strategy**.

Comau Change Model

The model is at the basis of all Comau's internal and external initiatives, both business and educational ones. It is divided into four phases:

1. *Active listening* to all the actors involved;
2. *Diagnosis* of the existing situation, of its criticalities and opportunities;
3. *Modeling* of the solution, co-designed with all project stakeholders, making extensive use of feedback and best practices, with a view to continuous improvement;
4. *Creation* of concrete tools with which to implement the solution through an iterative process of involving and listening to all the actors.

Figure 2. Comau change model

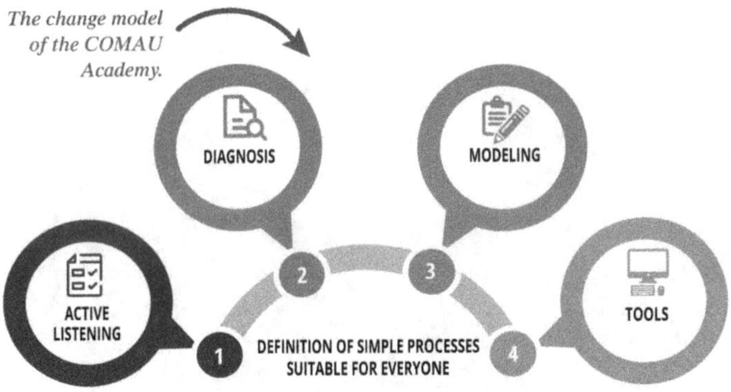

Comau Learning Model

As regards the training and educational initiatives carried out by the Comau Academy, the methodological reference is the proprietary learning model. According to the model, an effective initiative must offer the conditions to activate three dimensions, suitably modulated and regulated:

- *Relation*: people learn through interaction and social identification;
- *Content*: the learning process has a cognitive dimension;
- *Action*: learning develops mainly through action.

These three dimensions relate to the theoretical constructs linked to the conception of **learning as a social process**, situated and linked to competent participation in practices and operational activities. Knowing is a dynamic process (Lanzara, 2006) that occurs within a field of meanings that is activated within a system of activities (Engestrõm, 1987).

Figure 3. Comau learning model

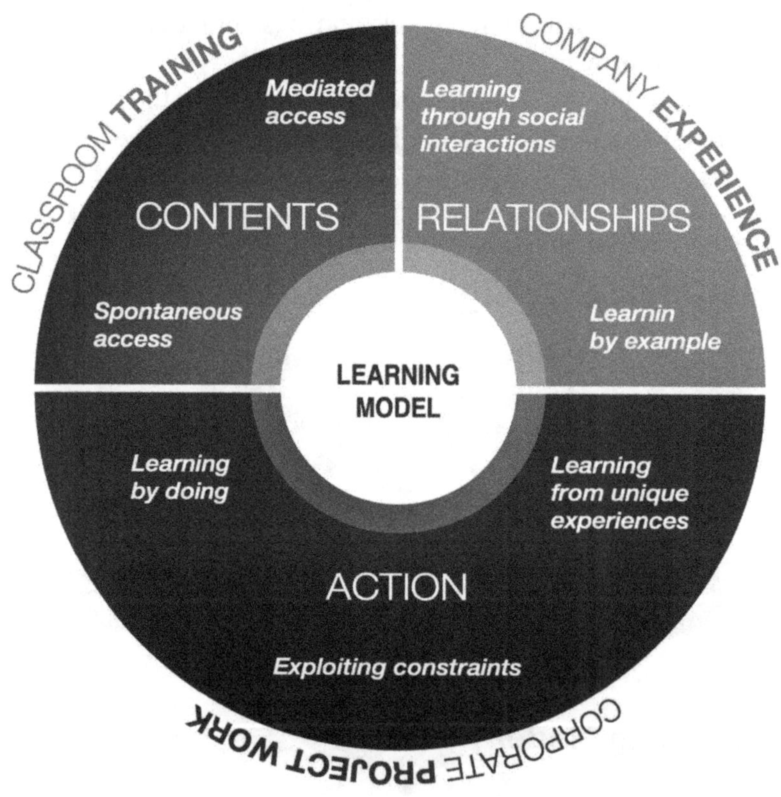

Partnership Strategy

Finally, there is a third distinctive element that connotes all the initiatives of the Academy, namely the fact that each of them is conceived in a synergistic way by three complementary actors: the world of work, the world of school and young people. The most effective learning situations, capable of responding to the challenges of the world of work of the future, are therefore the result of a holistic vision, the consequence of the comparison and integration that continuously occurs between the three actors. (Pinto et al., 2018). The Partnership Strategy carried out by the Comau Academy allows true interaction and contamination, capable of replicating the experiential dynamics present within the complex reality in which organizations and companies move today. It gives life to **new subjects, the ecosystems**. “An ecosystem of educational subjects is made up of actors who have a shared vision with respect to the changes taking place in the world of work and the need to disseminate and promote the development of knowledge different from traditional ones, which are more suitable and functional to the new context where we are located. (…) The scope of action of these new ecosystems expands and involves not only the world of school and the world of work, but also contaminates the organizational realities and people’s ways of working and interacting, focusing on values such as sustainability and social utility.” (Pinto et al., 2021)

The vision of the Comau Academy is in line with what emerges from the scientific literature, which emphasizes the **key role of corporate academies** in fostering the birth of new connections, capable of relating “different subjects” - young people, schools, universities and companies - and “new learning objects” - technologies, curricular disciplines, capabilities and work culture (Bellavista et al., 2016; Bersin, 2019).

E.DO EXPERIENCE, THE EDUCATIONAL ROBOTICS PROJECT

The drive to deal with strategic issues and skills for the workers and citizens of the future, through educational initiatives “from kids to professionals” is the way in which the Comau Academy is committed to **creating value for the people, the territory, and the ecosystem** in which it is immersed in terms of commons and generativity. In line with this vision, in 2016 the Comau Academy launched the Educational Robotic Project *e.DO Experience* with which it has reached an increasingly large and transversal audience.

Authors decided to focus their attention on this initiative because of the first interesting evidence that emerged in the previous chapters from desk-based research activities:

- Organizations can create **shared value in a generative and community-oriented key** developing values, knowledge and skills by means of **educational programs** that place **new technologies** at the center;

- **Educational Robotic initiatives** can be understood as a form of creation of shared social value;
- Value generated by educational robotics initiatives carried out by companies with a high technological impact can promote the **development of skills, knowledge and behaviors** useful to facilitate the entry into the 4.0 world of work.

The *e.DO Experience* Project is the result of **the deep interweaving between a technological dimension** - the educational robot e.DO - **and the world of teaching** - the *e.DO Experiences* that allow its use (Fregnan et al., 2020).

e.DO Robot

e.DO is a modular and open source, flexible and interactive **educational robot**, developed by Comau to **stimulate the creativity and active participation** of students during learning or to **support anyone who wants to explore the world of robotics** in depth, acquiring skills in using or programming a robot in a new and funny way. e.DO was initially developed for the world of manufacturers, technicians and robotics enthusiasts by a group of young technology passionate and experienced managers, who wanted to create **a modular and open-source robot** using a simple "open" technology, suitable to particularly interest the new generations: it is **an all-Italian story of innovation**, which today is achieving great success both in Italy and internationally.

Box 1.

e.DO is a small, open-source, multi-axis robot with built-in intelligence. It has six motorized axes and a maximum payload of 1 kg. The control logic is based on ROS (Robot Operating System). Users can create and execute programs to make e.DO perform specific activities such as movements and applications in space (from simple Pick & Place operations to managing more complex automated activities). Since it uses an open platform, e.DO can be programmed with a wide range of languages (such as C / C ++, Python or Java) and can interact with any ROS compatible hardware or software component. It is a suitable tool for students of all ages, but it is also of great interest to the most advanced users, who can directly enter the integrated control system, modify it and improve it.

Right from the start, e.DO proved to be a versatile and transversal tool: it is **like an industrial robot**, from a programming point of view, and it is **really user friendly** to use. It is also an **extremely secure product**, and this is of utmost importance in the context of educational applications. These

Figure 4. e.Do Robot

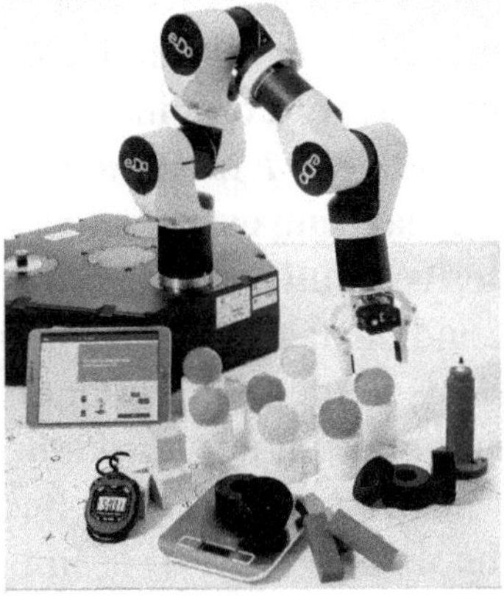

strengths were immediately noticed by the Comau Academy, that understood the **potential of e.DO as an educational robot** useful for learning the ABCs of robotics, to make school, extra-curricular learning, as well as the creative exploration of robotics and coding, more engaging and interactive. The e.DO project began to develop in 2016 thanks to the "*Robo-Scuola*" itinerant training initiative, carried out in Piedmont with the collaboration of the CRT Foundation. The initiative involved 3,000 students between the ages of 6 and 19, about 100 teachers and 38 schools, including primary and lower secondary and upper secondary schools. The initiative consisted of a series of laboratories in which e.DO served as a **didactic tool to make the understanding of normal subjects**, including mathematics and art, more intuitive and fascinating.

In 2017, the Agnelli Foundation inaugurated "COMBO" at its headquarters in Turin. This is **the first educational laboratory dedicated to STEM, robotics and coding** focused on the use of e.DO. From 2017 to 2020, this innovative educational space welcomed 7,600 students and 700 teachers from 240 primary and secondary schools in Turin, Piedmont, and other Italian regions. These two initiatives prompted the Comau Academy to give life to the *e.DO Experience*.

e.DO Experience

The *e.DO Experience* is an **educational platform developed for innovative learning methods** of STEM (Science, Technology, Engineering and Mathematics) subjects and for extra-curricular activities, with the aim of encouraging the **development of transversal skills** and **involving participants more directly** in training activities in which e.DO is the protagonist.

Like all Comau Academy education initiatives, it is based on the Comau Learning Model and it effectively combines the **dimensions of action, relationship and content**. The *e.DO Experience* educational platform offers an **original and alternative learning experience**. It is a constantly evolving reality that caters to **students of all ages**, from elementary schools to the university of the third age, and of all levels of training, from the professionals who want to renew their skills in the technological field, to the unemployed who seek to expand their opportunities through training courses, to the university researchers who need an adequate robotic platform to interface with all the most innovative technologies (AR, AI) or simulation and programming environments of the moment (Python, Matlab). Each *e.DO Experience* is designed to teach and learn STEM, Robotics and Coding subjects, choosing between different modes, duration and difficulty levels. Currently, there are three activated experiences.

e.DO Learning Center

Educational environments set up in foundations, schools, universities and training institutes. Participants, aged between 8 and 19, discover and experience e.DO in an interactive 4-hour session where STEM, Soft Skill and Work Culture subjects are integrated, which form the basis for their future growth. The Comau Academy is responsible for training the facilitators present in the centers, who manage the lessons. Each e.DO Learning Center has 4 or 5 e.DO robots and makes the necessary kits and teaching material available to both participants and facilitators.

e.DO Learning Lab

Educational programs dedicated to primary and secondary school teachers. Teachers receive the tools to manage an Educational Robotics laboratory lasting 2 hours in complete autonomy. Thanks to the use of e.DO robots, students between the ages of 8 and 19 can learn subjects such as Mathematics, Robotics and Coding in an innovative, practical and engaging way. Each e.DO Learning Lab requires the use of a maximum of 5 e.DOs or the virtual e.DO viewer.

e.DO Robotic Licence

Educational path dedicated to secondary school students. Participants learn to program and control e.DO through a desk simulator, simulating the industrial environment and acquiring robotics skills. High school students can become robotics experts while still in school, thus beginning to "taste" the world of industry and acquiring an internationally recognized Certificate in Robotics, of great value for manufacturing companies. To date, in Italy, 48 schools use e.DO to issue the Robotics License, which at the end of 2020 saw the participation of over 8,000 students.

In Italy there are currently 5 e.DO Learning Labs and 12 e.DO Learning Centers, but the *e.DO Experiences* are spreading all over the world. A Temporary Learning Center was created in Sweden, a hybrid training center with e.DO stations for the Stellantis headquarters in Belo Horizonte in Brazil, and permanent e.DO Learning Centers were created at training institutions in Russia and Brazil. Since 2019, e.DO Learning Labs have been acquired in Kazakhstan, India, Austria and the United States. Many e.DO robots have been sold in Poland, in support to the training packages that makes it possible to obtain the Robotics License.

e.DO Setting and Examples

The traditional setting in which the *e.DO Experience* takes place, namely the Learning Center, involves the **physical presence of participants and facilitators in a classroom**. In this space there are several tables on which, positioned on a dashboard, there is e.DO robot. The students, divided into **groups**, sit around the tables to carry out **practical activities** supervised by the facilitator.

The lessons take place within the robotics lab and last **four hours**. Within the setting it is possible to find, transversally to the different initiatives, the following elements:

- **Facilitator table**: table on which there is the teaching material; it is not part of the lab and only the facilitator has access to it.
- **Class table**: table on which are placed the educational objects useful during the activities; it is an integral part of the laboratory and can be accessed by the facilitator and the students (when indicated by the facilitator).

- **Work area e.DO**: table on which rest the dashboard and e.DO.
- **Group work area**: table around which the group sits, on which they can take notes, use the tablet, and perform calculations.
- **Learning island**: composed of the e.DO work area and the group work area.
- **Trunk**: box with learning materials, assigned to the group. The trunk is located under the table with the e.DO work area.

Depending on the type of lesson, the setting can commonly be enhanced by the following elements:

- **e.DO Cartesian plane**: laminated A3 Cartesian plane used for drawing with e.DO.
- **Team Cartesian plane**: plastic-coated A3 Cartesian plane used for drawing by hand.

Figure 5. e.Do setting

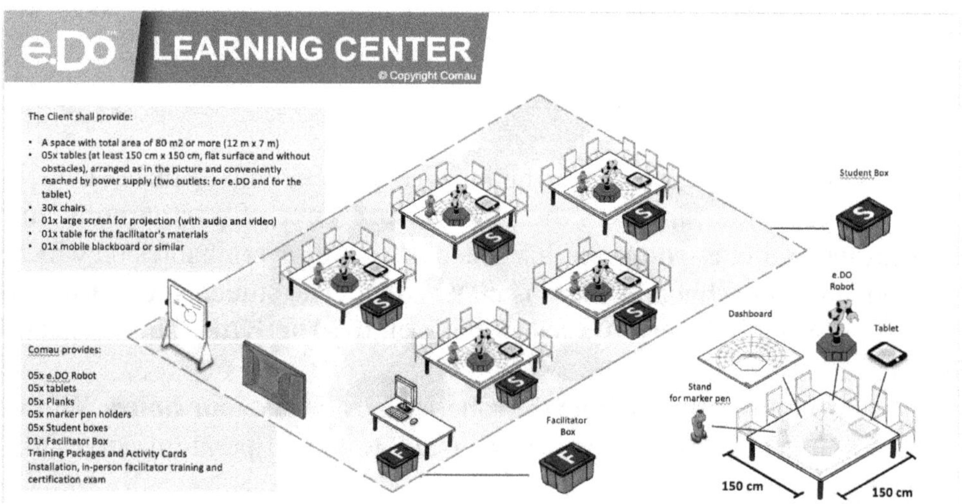

The first activity involves the **assembly of the components** and, therefore, the **assembly of e.DO**, which will be placed on the table in a vertical position to be then controlled through a tablet.

Figure 6. The assembly of e.DO

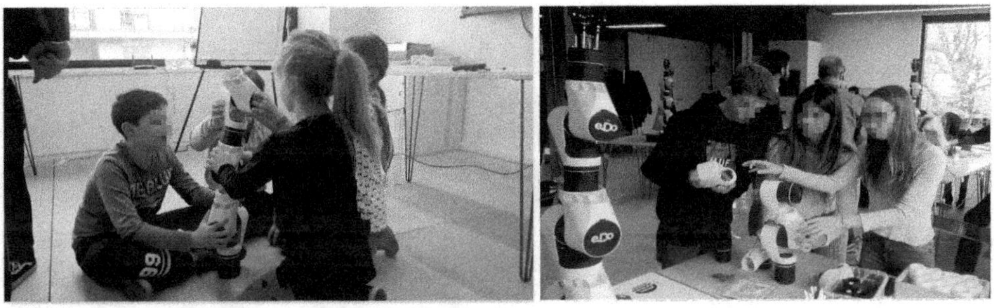

Math Examples

The students then carry out a series of **practical experiences aimed at learning curricular concepts** such as mathematics. An example of this application is the "Math in the Workshop" exercise, an activity aimed at learning the commutative property of addition by stacking tires of different thicknesses on a vertical rod, so as to meet specific dimensional constraints (e.g., "load the tires on the truck to create a stack of the required height"). Another typical activity in shop math involves moving these tires, carrying them on a scale to measure their weight (ex: "load the tires so that the truck does not exceed the maximum weight allowed by the scale on the ecology island"). These examples are calibrated mainly on **elementary school students**, and their usefulness is mainly to accompany students towards an **initial familiarization with the basics of mathematics**.

On the other hand, there are exercises of greater complexity, aimed, for example, at **secondary school students**. One of these activities is aimed at learning series and parallel production models through a collaborative module. Students are asked to use the scientific method to analyze the two production models and choose which of the two is best suited in a specific situation to be more effective and efficient. (e.g., "divided into roles, the goal of our food distribution center is to deliver 8 packages in 15 minutes.").

Another example of use for math familiarization for secondary students is the "Cartesian geometry for a startup" exercise. Here students are asked to use e.DO, in this case equipped with a marker pen, to identify the location of certain points and represent them on a Cartesian plane (e.g., "identify the location of supplier factories on the European map that will ensure that the greatest amount of material is recovered while minimizing the distance traveled"). Or, always in this exercise, students use e.DO to retrieve some objects and

Figure 7. "Math in the Workshop" activity

Figure 8. "Production Models" activity

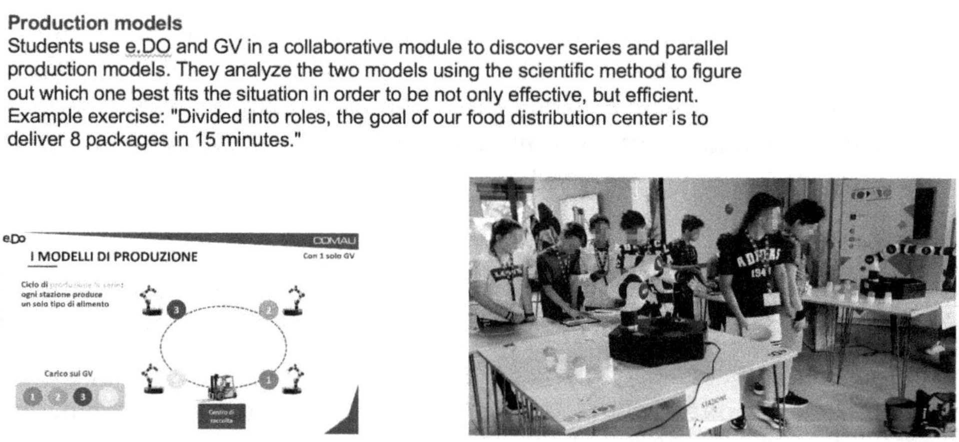

transport them on a scale, with the aim of discovering the physical relationship between mass and height of the moved objects and finally discovering their density (e.g. 1: "find the mathematical relationship that links the mass m and height h of the tires and represent it on a Cartesian plane."; e.g. 2: "represent the business plan and find the break-even point for our startup").

So, you can see how these exercises are designed to ensure students to **learn mathematical notions of increasing complexity**, starting with the concepts of height, weight, then moving on to the commutative property, the scientific method, the Cartesian plane, and linear physical relationships.

Figure 9. "Cartesian geometry for a startup" activity

Points on the Cartesian plane
Students use e.DO, equipped with a marker, to identify the location of certain points and represent them on the Cartesian plane.
Example exercise: "Identify the location of supplier plants on the European map that will ensure that the greatest amount of material is recovered while minimizing the distance traveled."

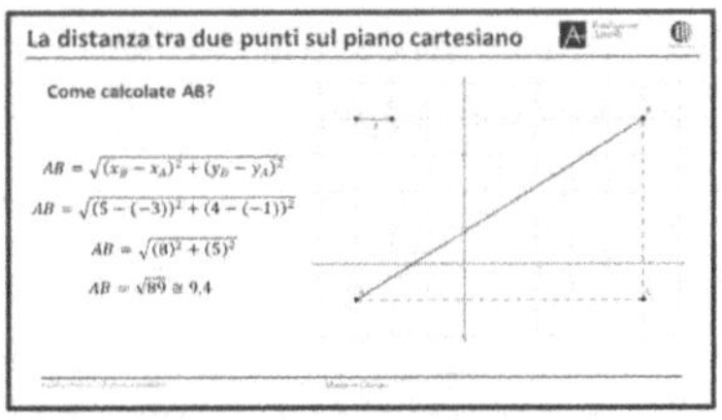

The Facilitator's Role

The activities carried out by the facilitator are aimed at a **functional accompaniment of students**, guiding the proper conduct of activities and the use of e.DO. The facilitator is also essential to **introduce participants in the logic of learning** typical of the Learning Center, no longer based on the classic model of frontal lessons, but supported by principles of **experiential learning**, active, engaging and with a strong emotional impact.

The activities carried out by the facilitator can be summarized as follows:

- **Welcome**: The facilitator introduces himself, illustrates the pivotal theme of robotics, divides students into groups and directs them to their respective area of group work. Finally, he/she introduces the learning module and presents the workshop agenda;
- **Lab Rules**: Facilitator explains lab safety rules (emergency exits, ...) and shows safety video with e.DO;
- **Introduction to Robotics**: the facilitator shows a video with a robot in action, asks students for a definition of robot and then builds a shared one, explaining the etymology of the word. He also brings examples of robots that are now part of our daily lives and analyzes the differences between robots and humans;
- **e.DO components**: the facilitator introduces e.DO and describes its components, namely the egostaffe, the joints, the gripper, the control unit, the emergency button and the tablet;

- **Assembly of e.DO**: each group has 10 minutes to assemble e.DO, this activity is done in groups and on a horizontal plane. The different components are connected but not interlocked. In this phase, the facilitator supports the students and, once this first task is completed, assesses the results;
- **Handling**: the facilitator explains to the participants how to use the interface for handling e.DO, the students try to use the commands under his supervision;
- **Work Metaphor**: the facilitator introduces the work metaphor that will guide the day's activities (e.g., tire shop);
- **Task Explanation and Activity**: the facilitator explains the single activity to be performed by the students (e.g. tire shop: "Inside our shop we want to create a pile of tires, of a precise height"), illustrates the rules and objectives. After that, once the task has started, the facilitator moves between the different groups to monitor the activities and support the students. Once the task is completed, the facilitator evaluates the groups' activities.
- **Lesson**: the facilitator explains the theoretical and curricular content that supported the development of the task (e.g., commutative, associative, and dissociative properties).
- **Lesson Learned**: the facilitator summarizes what has been done and the main themes, robotics and mathematics, addressed. He/she then emphasizes the importance of the practical use of the mathematics that is studied in School (with examples).
- **Task Verification**: the facilitator discusses with the participants, at the end of each task, about the achievement of the results.
- **Feedback and Conclusion**: the facilitator provides feedback to participants on the activities carried out within the workshop, encourages students to do the same and ask questions. At this time, reflections and reasoning about the experience are shared. At this point the facilitator thanks and greets the class.
- **Unpacking**: the facilitator turns off the tablets and puts them away in a safe place, turns off e.DO and rearranges the materials within the trunk they belong to, the group work areas, the teaching island and the facilitator's table.

The facilitator is a **central figure for the *e.DO Experience***, his contribution is in fact fundamental for the accompaniment of students, not only in learning and in the execution of the task but also and above all for understanding the experiential, active, involving and strongly emotional logic that guides the training activity.

Educational Products

Within the *e.DO Experience* project, some more **educational tools** have been developed to accompany students and teachers to discover truly unique didactic experiences: Activity Cards, didAPPs and Training Packages.

Activity Cards

Educational cards designed to accompany students and teachers during the educational experiences conducted with e.DO robot and e.DO Cube. They are a condensation of theory, practical exercises and learning tests collected in a booklet divided into three different sections of work. Comau Academy has created about 300 Activity Cards, dedicated to the study and the deepening of specific topics in the fields of Mathematics, Physics, Technology, Robotics and Coding. The cards are diversified by subject and learning level, and teachers can integrate them into their lessons.

didAPPs

2-hour interactive educational paths developed by the Comau Academy to guide teachers and students in every phase of the learning path with e.DO robot and e.DO Cube. These are complete didactic modules consisting of video lessons, theoretical and practical in-depth materials, a virtual e.DO viewer, verification tests and a guide to support teachers in every moment of the training project. They are designed to integrate disciplines such as Robotics, Mathematics, Coding with specific transversal skills (Teamwork, Problem Solving, Creativity). Designed as an extremely flexible tool, they allow teachers to independently choose the topic and the number of didAPPs to use, based on the study path they intend to pursue with their students.

Training Packages

Complete workshops of 4 hours each that allow guided teaching of STEM subjects, robotics and programming with e.DO. These are complete and turnkey educational projects to be conducted with the support of e.DO and e.DO Cube, including didactic material and video lessons, differentiated by learning level and age groups. Enriched with theoretical and particular in-depth material, final verification tests and presentations of interesting use cases, they offer a truly immersive experience within the educational world of e.DO.

Thanks to these educational products, the Comau Academy offers students and teachers the opportunity to take advantage of over 370 hours of training with e.DO, in the classroom or in distance learning.

e.DO Cube

In 2019, the Covid-19 pandemic largely interrupted face-to-face training activities, replacing them with Distance Learning (DAD). New training needs emerged. Moving teaching activities to a virtual environment required teachers and students to make a great effort to adapt to new tools, skills and contents. We have witnessed a greater digitization of the educational system. To cope with this unprecedented scenario, the Comau Academy also had to **rethink the *e.DO Experience* in a fully digital key**. To this end, it quickly developed **e.DO Cube, a digital robot** created specifically for distance learning. e.DO Cube is a "Digital Twin", **a virtual twin of the e.DO robot**, which allows students and teachers to carry out online the same didactic programs that can be carried out in person with the physical robot.

Box 2.

e.DO Cube is a digital device that uses the same hardware and software as the e.DO robot.
Connected to a PC or tablet, via USB cable, it allows you to view a virtual robot on the screen in a Cartesian space, where it can be guided to move, pick up or move objects, performing the same operations as a real e.DO. When programmed and moved within the digital platform, e.DO Cube allows you to learn and teach, even remotely, STEM subjects, robotics and coding, without teachers and students having to give up both the effectiveness of the teaching experience offered from practical lessons and the sharing of information interacting with the working group.

During the lesson the teacher uses e.DO robots in the classroom, while the students, from home, through the e.DO Cube that they view on their computers, can perform the assigned tasks and send them to the teacher. The programs created by the students can thus be transferred from the e.DO Cube box to the real e.DO to concretely test their effectiveness. e.DO Cube is therefore **a valuable tool both for distance learning and for mixed learning**. Thanks to the potential of the e.DO Cube box, for example, it is possible to activate

Figure 10. e.DO Cube

synchronous and asynchronous collaboration projects. Students can work in groups within virtual rooms - shared online environments - where participants can remotely contribute to carry out common projects.

e.DO Cube Setting and Examples

e.DO Cube is a digital device that uses the same hardware and software of e.DO robot. Connected to a pc or a tablet, through a USB cable, it allows to visualize on the screen a virtual robot in a Cartesian space, where it can be guided to move, pick up or move objects, performing the same operations of a real e.DO. When programmed and moved within the digital platform, e.DO Cube allows STEM subjects, robotics and coding to be learned and taught, even at a distance, without teachers and students having to give up both the effectiveness of the teaching experience offered by hands-on lessons, and the sharing of information and interaction with the workgroup.

A typical setting for implementing educational robotics initiatives that take advantage of the potential of e.Do Cube takes the form of this type of environment.

The experience takes place in a computer room or other equipped classroom in the school. The students are divided into small groups and each group has at least one computer to view and operate their virtual robot - e.DO Cube - and take part in the proposed activities. The classroom is equipped with a projector/LIM and multimedia audio/video/microphone system, clearly visible

Figure 11. e.DO Cube virtual simulator

Figure 12. e.DO Cube setting

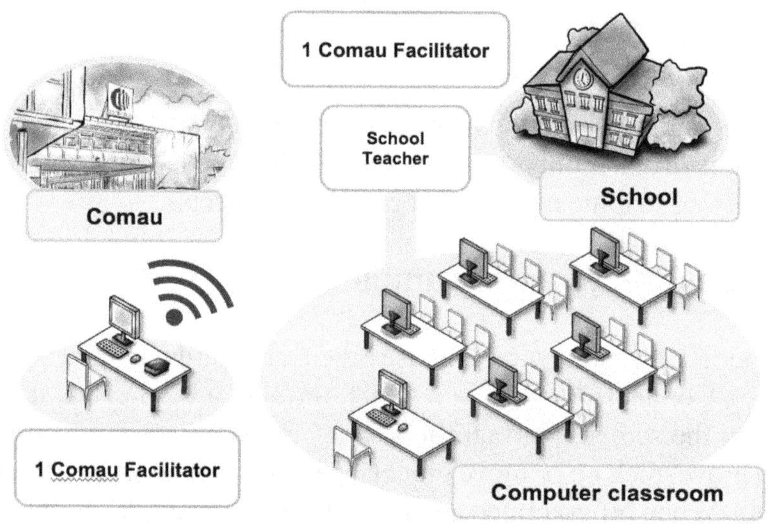

and audible to the students, to allow the Comau tutor to show the slides and interact with the classroom.

The workshop sees the presence of a Comau facilitator who manages the activities at a distance and a school teacher who supports the students in presence.

RoboSpace

Let's analyze an example of an educational robotics initiative conducted thanks to e.DO Cube. RoboSpazio is a project of Comau and Fondazione Agnelli.

It is an educational workshop that takes place in the educational laboratory dedicated to STEM, robotics and coding, with the support of e.DO Cube.

Students, guided by their teachers and followed remotely by a tutor, use e.DO Cube to overcome, in groups, math and robotics challenges and problems whose solution requires specific levels of competence in the subjects involved. The activity aims to improve the understanding of space and Cartesian reference systems, getting to understand the basics of global positioning systems, such as GPS.

Figure 13. RoboSpazio educational objectives

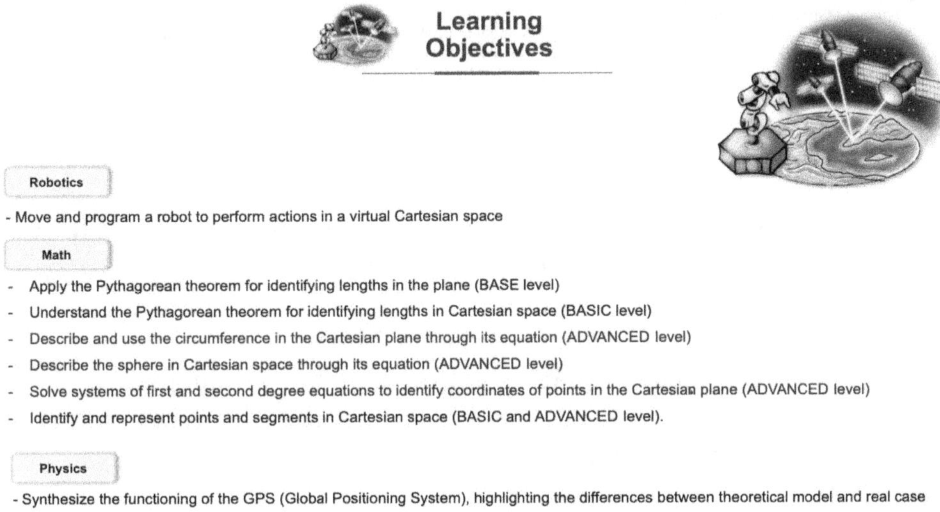

The workshop lasts 4 hours. All secondary school classes, of every course and address, from every Italian location can participate. During the workshop hours students:

- are guided in class by their teacher and remotely by a COMAU trainer in an adventurous race against time to shed light on a mysterious disappearance;
- take part in a series of challenges, to see how the math and physics topics from the school curriculum are useful and applied in everyday life.

Figure 14. RoboSpazio structure

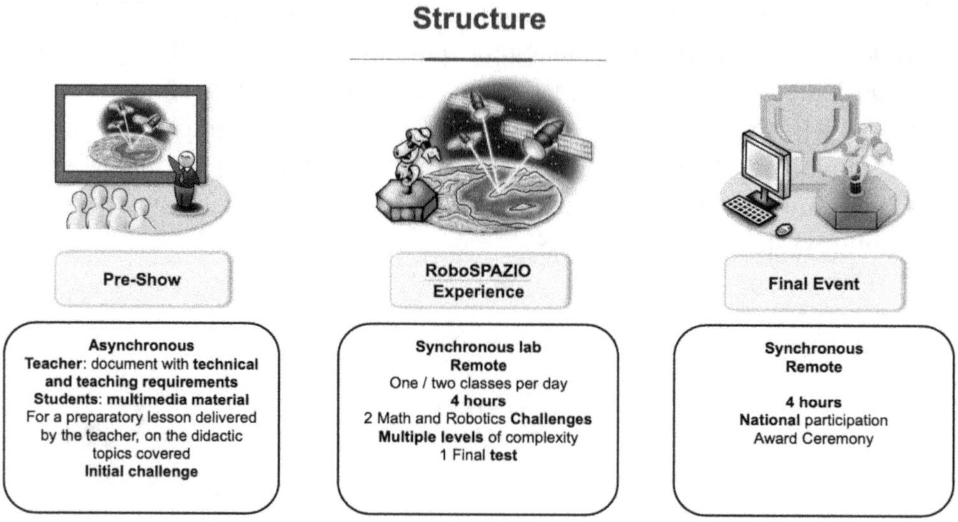

The activity is divided into three macro-blocks of Robotics and Math and GPS (one hour each) each divided into 3 phases:

- frontal component, focused on expanding knowledge
- activation phase: launch and resolution of challenges in groups
- correction and feedback: analysis of what has been done and sharing of the lesson learned among all students.

Figure 15. RoboSpazio agenda

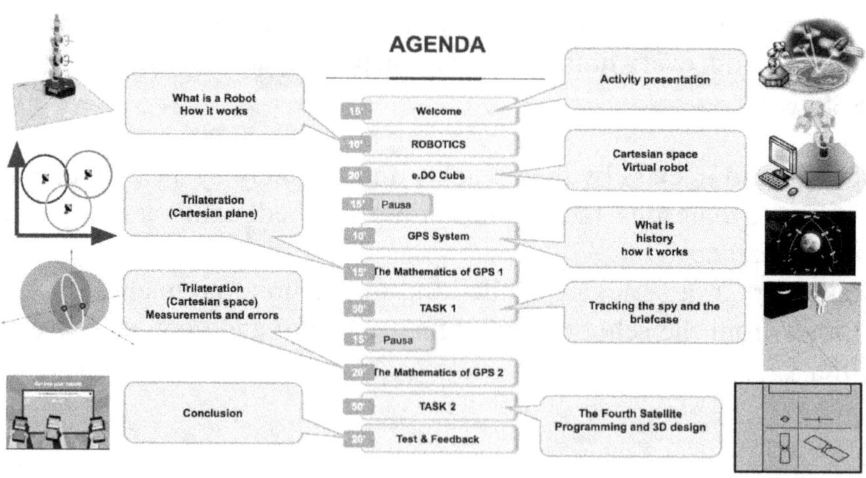

The workshop is available in three different levels of complexity, to adapt the experience to the age and curriculum of the different participating classes.

- **Level 1**
  - Reference systems and the Cartesian plane;
  - Identification and representation of points and segments through the use of coordinates in the Cartesian plane
  - The Pythagorean theorem;
  - Application of the Pythagorean theorem for the identification of lengths from Cartesian coordinates.
- **Level 2**
  - Reference systems, the Cartesian plane and Cartesian space;
  - Identify and represent points and segments through the use of coordinates in the plane and in Cartesian space;
  - Describing the circumference on the Cartesian plane through its equation;
  - The position of two circumferences in the Cartesian plane;
  - Methods of solving systems of equations for identifying coordinates and points in the Cartesian plane
  - Calculation of uniform motion;
  - Errors of measurement and uncertainty of measurement.
- **Level 3**
  - Reference systems, the Cartesian plane and Cartesian space;
  - Identify and represent points and segments through the use of coordinates in the plane and in Cartesian space;
  - Describing the circumference on the Cartesian plane through its equation;
  - The position of two circumferences in the Cartesian plane;
  - Methods of solving systems of equations for identifying coordinates and points in the Cartesian plane;
  - Describing the sphere in the Cartesian space through its equation;
  - Calculation of uniform motion;
  - Measurement errors and uncertainty of measurement;
  - Relativity of space and time;s
  - General and restricted relativity.

The most interesting features of this educational robotics initiative include:

Figure 16. RoboSpazio workshops

- the use of **robotics as an innovative learning tool**; students have the opportunity to control and program a virtual robotic arm to perform movements in a three-dimensional Cartesian space, directly from their computer;
- the adoption of a **hybrid solution**, which takes advantage of the possibilities offered by the digital twin of e.DO - e.DO cube - to allow students to experience different types of dynamics, encouraging the learning of digital skills also in coherence with the dynamics typical of the outside world, increasingly characterized by hybrid ways of collaboration and work;
- the **subdivision into teams**; to stimulate students to collaboration and accustom them to team work;
- the **elimination of geographical constraints**; classes from all over Italy can participate in the experience, regardless of where they are located;
- the **gamification**, which is played out on two levels: the transformation of didactic topics into **engaging challenges** and the **competition at national level**. While participating in the workshop, the classes, based on the results achieved during the resolution of the challenges, receive a score and are included in a national ranking. At the end of the school year, those who have reached the top of the podium will have the opportunity to participate in a final challenge (also remotely).

- the **role of the teacher**; the lead teacher (preferably a math, physics or other STEM teacher) is required to have skills in managing the necessary technologies, as well as active support during the lab.

To summarize, below is a table comparing **similarities and differences between e.DO and e.DO Cube**, compared to: role of the facilitator, use of space, prompted activities, objectives.

Table 1. e.Do and e.DO Cube: similarities and differences

	e.DO	e.DO Cube
ROLE OF TUTOR	The tutor helps the participants to trigger a self-induced growth of their knowledge.	
	The tutor is physically present in the lab with the participants. He/she autonomously manages every type of moment during the activity.	The tutor is NOT physically present in the lab with the participants but he/she always interact with them through audio/video support. He/she needs support from a teacher present in the lab, to guide the students during the activity.
USE OF SPACE	For the activity of a class (around 25-30 students) a room with 5 stations with 1 e.DO each is used (5-6 students each e.DO). The students have to reach this **robotics lab**. **e.DO is a REAL OBJECT.** e.DO is always located on a table and the robot is controlled by a tablet.	For the activity of a class (around 25-30 students) a computer lab with 10 computers and 1 IWB is used (2-3 students each e.DO Cube). The students can join by any **computer lab**. **e.DO Cube is a VIRTUAL ENVIROMENT.** e.DO Cube is visualized and controlled on the screen of a laptop/computer.
PROMPTED ACTIVITY	The activities are designed to be carried out in small groups, to stimulate cooperation, to develop soft skills and participants can be challenged in the resolution of various tasks. It is possible to program the robot in order to perform pick and place activities, with different programming languages and interfaces (step-by-step programming, Blockly and visual programming languages, Python and textual programming).	
	It is easier to understand the tridimensional space. Due to the presence of more than one e.DO station, it is possible to deliver cooperative activities based on collaboration of various teams and robots. It is more engaging and immediate.	It is possible to draw in a 3D space, allowing to develop concepts related to abstraction skills. Due to its virtual nature it is possible to manage activities with teams connected by different locations.
OBJECTIVES	e.DO / e.DO Cube have been developed to teach STEM contents to students using Robotics as starting point.	

Teacher and Student Perspective

The authors decided to look more closely at the **views of teachers and students** in order to gain further insights into the *e.Do Experience*. This is of particular importance because most research into classroom robotics are external to the curriculum, **whereas the *eDo Experience* actually integrates robotics into the curriculum**.

Authors focused their attention on the **Combo project** managed by Comau and Fondazione Agnelli, the first educational laboratory dedicated to STEM, robotics and coding focused on the use of e.DO.

In the **2017-2018 school year**, the numbers for the Combo project were:

- 2446 students (997 primary school students, 535 lower secondary school students, 914 upper secondary school students)
- 211 teachers (104 primary school teachers, 55 lower secondary school teachers, 52 upper secondary school teachers)
- 114 classes
- 448 hours of workshops
- 3 teaching modules

In the **2019-2020 school** year, the numbers for the Combo project were:

- 2025 students (651 primary school students, 793 lower secondary school students, 581 upper secondary school students)
- 197 docenti (75 primary school teachers, 78 lower secondary school teachers, 44 upper secondary school teachers)
- 99 classes
- 396 hours of workshops
- 4 teaching modules

Teacher Perspective

Starting from **teacher perspective**, the research examined feedback given by them at the end of the workshops in which they participated.

From a **qualitative point of view**, teachers expressed positive comments. Here you can read some verbalizations.

Primary School

- *The balanced and well-structured learning path perfectly reaches the objectives and highlights how important it is to have skills and not only knowledge.*
- *Something different, close to the interests of children.*
- *Great alternative method to show math in a concrete and intuitive way.*
- *Very useful for children with difficulties, it would be very nice to use other activities / teaching hours. It is important to focus on key concepts and explain them slowly. Beautiful! My kids are difficult! Congratulations!*
- *Innovative teaching experience that develops observation and problem solving.*

- *Excellent management of the activity, clear and simple language, a well-defined structure that still left room for action and experimentation for the children. action and experimentation to the children. Thank you!*
- *Group work to deepen and enrich rigorously the contents of mathematics.*
- *Acquired knowledge has become "Acted Skills".*
- *Interesting - Stimulating.*
- *It would be good to have the possibility of a few more structured meetings with different topics. Thank you very much!*
- *Effective in that learning comes from practical experience and fun.*
- *Exciting experience that has involved the whole class group, managing to reinforce and anticipate mathematical topics.*

Lower Secondary School

- *A fantastic opportunity, an eventual steppingstone for the future of school and work.*
- *It was a unique experience to learn about working methodologies from the world of "grown-ups". They have had the opportunity to put into practice concepts seen in class, "scientific method" and mathematics by applying them to the productive world.*
- *Exciting and stimulating.*
- *The activity also allowed the students to express practical skills.*
- *Involving, reflective, correct for the age of the students.*
- *To be repeated again.*
- *Exciting - Innovative.*
- *The proposed activity allows children to learn about technology and innovation in simple and effective language.*
- *Excellent experience.*
- *Stimulating update on new technologies applicable to teaching.*
- *Teachers and students very involved, we thank you for the experience.*
- *Didactically innovative.*
- *Good involvement and good alternation between theory and practice.*

Upper Secondary School

- *A true didactic for competences, absolutely transversal and educational at 360.*

- *Extremely useful on the didactic plan and on that of the socialization of the boys. It helps kids understand possible applications of the concepts studied.*
- *I hope it stimulates the students to be more committed to the study of mathematics and especially its applications.*
- *Experience that breaks the routine of the lessons and opens the horizons.*
- *Very well organized. Congratulations.*
- *Interesting and highly engaging experience for the kids.*
- *Very useful through practice use of theory.*
- *Beautiful! You learn to collaborate and manage the group.*
- *It was an engaging and exciting experience, complete and full of ideas to be expanded in the educational offer of the school.*
- *Active participation.*
- *Excellent experience in involvement, application, and from an educational standpoint.*

From a **quantitative point of view**, teachers **confirmed the positive feedback**. The scale used to collect their ratings ranged from 1 to 4. Below the results collected in 2017-2018 and 2019-2020.

Figure 17. Quantitative feedback from teachers in 2017-2018

		Class number	Teacher number	TEACHER EVALUATION			
				E	F	G	H
Primary School		**47**	**104**	**4,00**	**3,95**	**4,00**	**3,98**
	III	14	32	4,00	3,97	4,00	4,00
	IV	13	29	4,00	3,97	4,00	3,97
	V	20	43	4,00	3,91	4,00	3,98
Lower Secondary School		**26**	**55**	**3,93**	**3,78**	**4,00**	**3,89**
	I	11	23	3,96	3,78	4,00	4,00
	II	11	24	3,92	3,71	4,00	3,75
	III	4	8	3,88	4,00	4,00	4,00
Upper Secondary School		**41**	**52**	**3,94**	**3,82**	**3,96**	**3,94**
	II	7	8	4,00	4,00	4,00	4,00
	III	17	23	3,91	3,83	3,91	3,87
	IV	17	21	3,95	3,75	4,00	4,00
TOTAL		**114**	**211**	**3,97**	**3,87**	**3,99**	**3,95**

E: How would you rate your students' appreciation of the activity?
F: How would you rate the degree to which the activity was useful to your students?
G: How would you rate the speaker in terms of clarity and competence?
H: How would you rate the facilitator's ability to engage students?

Figure 18. Quantitative feedback from teachers in 2019-2020

	Class number	Teacher number	TEACHER EVALUATION			
			E	F	G	H
Primary School	**31**	**68**	**3,90**	**3,93**	**3,97**	**3,96**
III	10	21	3,90	3,90	4,00	3,90
IV	12	25	3,88	3,96	3,96	3,96
V	9	22	3,91	3,91	3,95	4,00
Lower Secondary School	**32**	**68**	**3,87**	**3,87**	**3,99**	**3,94**
I	8	16	3,88	3,88	3,94	3,88
II	18	41	3,88	3,88	4,00	3,98
III	6	11	3,82	3,82	4,00	3,90
Upper Secondary School	**23**	**34**	**3,94**	**3,85**	**3,97**	**3,91**
II	10	15	4,00	3,93	4,00	3,93
III	6	7	4,00	3,71	3,86	3,86
IV	6	11	3,82	3,91	4,00	3,91
V	1	1	4,00	3,00	4,00	4,00
TOTAL	**86**	**170**	**3,9**	**3,89**	**3,98**	**3,94**

E: How would you rate your students' appreciation of the activity?
F: How would you rate the degree to which the activity was useful to your students?
G: How would you rate the speaker in terms of clarity and competence?
H: How would you rate the facilitator's ability to engage students?

Student Perspective

Analyzing feedback from **students** in schools that participated in the Combo project modules, the evidence shows **great engagement**. Students say they learned many new things and had fun doing it and appreciate the opportunity to interact with technological artifacts such as e.DO robot.

Here some verbalizations.

Primary School

- *I had a lot of fun, and it was a great learning experience.*
- *I had so much fun today and I think it's the best outing we've had with the class and I learned so much*
- *I had so much fun today and would like to do it again 1000 times I loved the quiz.*
- *I had a lot of fun and I realized that having fun you can learn. I was greeted with kindness and they were very nice. I hope to return!*
- *For me this lab was great, a mix of robotics and math. It is a lab worth doing.*

- *I enjoyed it so much and I hope they can build more robots in the world, and I realized that to program you need a team!*
- *I want a e.DO at home!*
- *It was beautiful, the most beautiful morning in history!*
- *I liked it a lot because I learned a lot about robots.*

Figure 19. Some feedback from primary school students

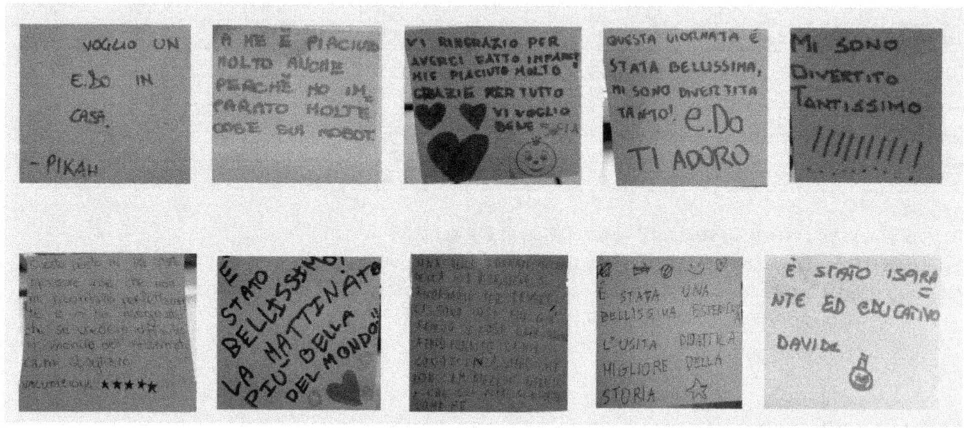

Lower Secondary School

- *I am very happy, and I wish all the classes were like this I would learn a lot more because I would be careful*
- *This activity was very interesting and engaging and thanks to the explanation while understanding difficult topics or otherwise not easy to understand we were able to work in groups respecting and enjoying ourselves.*
- *Simply fantastic!*
- *I had a lot of fun, I learned a lot, and the robots, which I never thought I would touch, opened up a world to me, I fell in love with robots.*
- *Today I learned the basics of work and from now on I love robots*
- *The workshop was fun and enjoyable thanks to the facilitator who was understanding and personable.*
- *I had so much fun today, I collaborated with my friends and here is a place where you have fun and learn technology. For me this lab is 5 stars!*

Figure 20. Some feedback from lower secondary school students

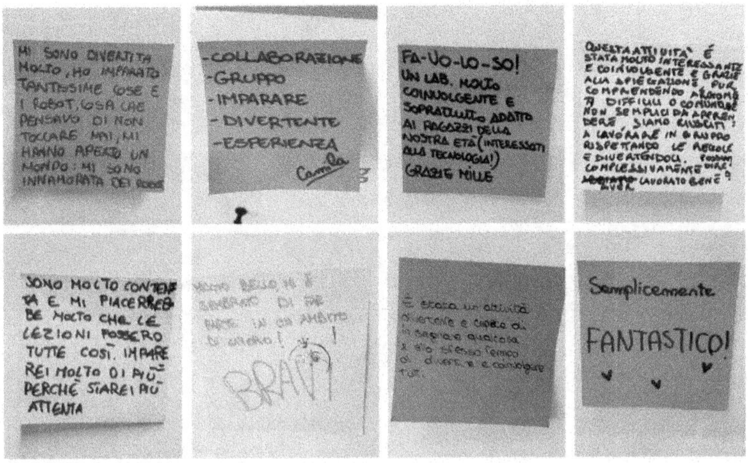

Upper Secondary School

- *Exciting, fun and innovative*
- *It could be one of my addresses for the future*
- *Super interesting look into the present and future*
- *Really exciting experience stimulates creativity, skills acquisition and putting yourself out there*
- *It made me realize that there are many ways to solve a problem, even different approach*
- *The topic is very interesting and the way it is explained is just perfect. in my opinion in fact this is the best way to learn things while having fun, with the involvement.*
- *Although I have already been there the happiness of coming here is very high*
- *The workshop was very engaging and fun because we worked in groups and used advanced technological tools with appropriate materials.*

From a **quantitative point of view**, also in this case students **confirmed the positive feedback**. The scale used to collect their ratings ranged from 1 to 4. Below the results collected in 2017-2018 and 2019-2020.

Figure 21. Some feedback from upper secondary school students

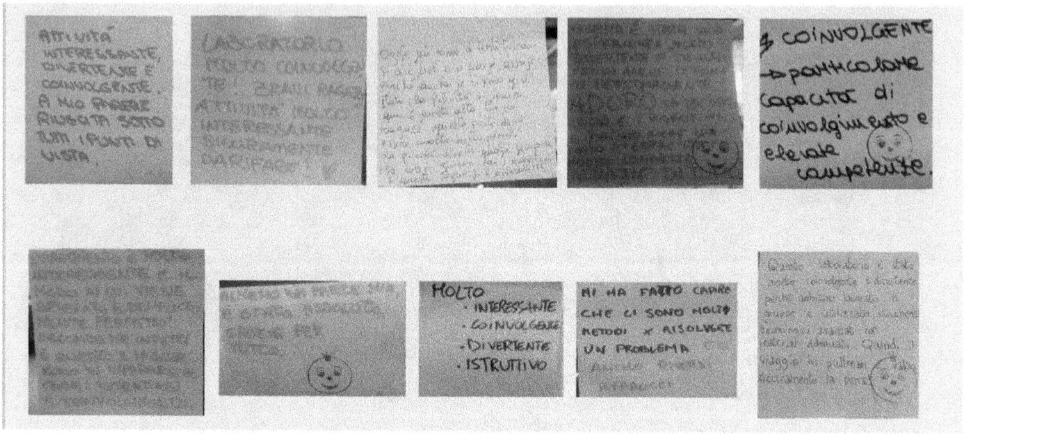

Figure 22. Quantitative feedback from students in 2017-2018

	Class number	Student number	STUDENT EVALUATION A	B	C	D
Primary School	**47**	**997**	**3,79**	**3,89**	**3,82**	**3,85**
III	14	285	3.86	3,88	3,80	3,85
IV	13	281	3,75	3,91	3,81	3,88
V	20	431	3,76	3,87	3,84	3,83
Lower Secondary School	**26**	**535**	**3,76**	**3,85**	**3,79**	**3,79**
I	11	245	3,75	3,86	3,78	3,77
II	11	216	3,75	3,84	3,81	3,79
III	4	74	3,84	3,88	3,81	3,84
Upper Secondary School	**41**	**914**	**3,34**	**3,74**	**3,81**	**3,78**
II	7	158	3,45	3,71	3,72	3,88
III	17	398	3,31	3,68	3,80	3,71
IV	17	358	3,34	3,82	3,85	3,82
TOTAL	**114**	**2446**	**3,62**	**3,82**	**3,81**	**3,81**

A: How much did you learn/review?
B: How much fun did you have?
C: Was the facilitator clear and knowledgeable?
D: Was the facilitator engaging?

CONCLUSION

The desk-based research has allowed to describe the innovative Educational Robotics project leaded by the Comau Academy: the *e.DO Experience*. Analyzing in detail the key elements that characterize the experience, the first

Figure 23. Quantitative feedback from students in 2019-2020

	Class number	Student number	STUDENT EVALUATION			
			A	B	C	D
Primary School	**31**	**591**	**3,60**	**3,81**	**3,62**	**3,66**
III	10	202	3,65	3,84	3,53	3,63
IV	12	216	3,59	3,78	3,65	3,68
V	9	173	3,57	3,79	3,67	3,68
Lower Secondary School	**32**	**694**	**3,61**	**3,72**	**3,62**	**3,59**
I	8	245	3,65	3,72	3,57	3,53
II	18	216	3,63	3,7	3,64	3,59
III	6	74	3,45	3,75	3,63	3,70
Upper Secondary School	**23**	**914**	**3,34**	**3,68**	**3,76**	**3,75**
II	10	206	3,34	3,73	3,79	3,77
III	6	131	3,41	3,69	3,83	3,76
IV	6	110	3,25	3,58	3,64	3,71
V	1	13	3,38	3,69	3,62	3,77
TOTAL	**86**	**1745**	**3,52**	**3,82**	**3,81**	**3,81**

A: How much did you learn/review?
B: How much fun did you have?
C: Was the facilitator clear and knowledgeable?
D: Was the facilitator engaging?

result obtained is the definition of the **main benefits for the Participants: teachers and students**.

Benefits for Teachers

- transferring notions and knowledge by activating new ways of learning and teaching
- enhancing theory with practice, with a glance at "gamification"
- encouraging participation and inclusion, facilitating interaction and learning even in the presence of cognitive difficulties (SLD, ADHD or BES)
- acting as a facilitator, not as a teacher
- adapting one's working method to new educational needs
- managing the class group in an easier way
- learning STEM, Robotics and Coding subjects

Benefits for Students

- learning in an original, creative and alternative way, compared to conventional methods

- familiarizing with the technology
- learning STEM, Robotics and Coding subjects
- understanding the potential of a simplified and intuitive visual language
- becoming "protagonists" of the laboratory activity
- learning through challenges and games
- developing and enhancing soft skills and transversal skills
- collaborating to obtain a common result in a stimulating way
- "tasting" the world of industry and the world of work

These benefits seem to be attributable to the areas of value that emerged during the desk-based research. In terms of **value creation**, Educational Robotics initiatives such as e.DO Experience seem to **serve several functions, particularly related to their effects on the Participants**: Skill Creation, Value for New Citizenship, Empowerment and Inspiration & Orienteering.

However, other interesting data emerge, which are reported below, and which highlight how **the *e.DO Experience* Project seems to have brought value not only to the participants, but also to the Company itself** (Comau and its Comau Academy) **and to the Territory** that hosts it.

- The "culture" and know-how that characterize the industrial system, when shared outside a company context, can enrich and enhance the world of education and the world of work;
- A well-constructed and articulated Educational Robotics initiative enables important business objectives to be achieved;
- The enhancement of intuitions and expertise of the company's people can generate valuable innovation;
- The important results obtained amplify the national and international reputation of the Company's brand;
- The innovative use of an educational robot favors the pragmatic approach to disciplinary learning;
- Robotics allows the reinforcement of soft skills, encouraging involvement, participation and inclusion;
- Educational Robotics is a powerful tool for innovation in the world of traditional teaching;
- Innovative Educational Robotics initiatives can help young people to become the real protagonists of the future;
- This new approach to teaching starts from robotics to encourage and enhance the learning of curricular subjects and business culture;

- Organizations that practice learning agility internally are more capable of teaching such key skill externally as well;
- Making innovative technologies and educational programs available is useful to allow a wide and diversified public to acquire the notions and "capabilities" necessary to be able to live and work in the new social, cultural and professional landscape, in continuous transformation, outlined by the Fourth Industrial Revolution;
- Expanding and strengthening a "new educational paradigm" is essential to create - in a logic of continuous collaboration - innovative teaching proposals to inspire and encourage the new generations to grow in a stimulating and effective way.

These evidences suggest that the *e.DO Experience* - a real project from a high-technological impact company – can bring to light **two additional levels of value creation** related to Educational Robotics initiatives. The Comau *e.DO Experience*, in fact, seems to direct the creation of value at participants level, but also at a company and territorial level.

The **Value for Company** seems to be related to the following areas of value:

- Core Business/Core Mission
- Brand (Reputation/Marketing)
- Employee Branding
- Customer Social Responsibility (formal obligations)
- Customer Social Responsibility (true commitment to create a better world)
- Open Innovation (get the innovation potential that is present outside the organization - talents, know how, ideas)
- Employees' Exposure

The **Value for Territory**, instead, seems to be related to the following areas of value:

- Creation and Dissemination of Culture
- Didactic Innovation
- Contribution to the Development of the Skills required from the World of Work
- Contribution to the Creation of New Ecosystems

The next chapter will **summarize all the results obtained from the desk-based research** activity to reach **the first two research goals**:

1. Describing the Digital Transformation scenario, the society and companies' evolution and the request for new skills
2. Identifying the specific values of educational robotics initiatives, also describing an innovative educational robotics project: the Comau *e.DO Experience.*

Afterwards, authors will **move on to the last three goals** of the research:

3. Verifying and exploring the values generated by the Comau *e.DO Experience* and to identify, list and describe the main indicators for each value
4. Measuring the weight of each value and its respective indicators in the perception of those who designed and delivered the Comau *e.DO Experience* educational robotics project
5. Exploring perceptions and representations of participants in the Comau *e.DO Experience* educational robotics project.

REFERENCES

Bellavista, M., & Faggin, A. (2016). *Management della formazione. Dal servizio formazione alle Academy/Corporate University*. Franco Angeli.

Belpaeme, T., Kennedy, J., Ramachandran, A., Scassellati, B., & Tanaka, F. (2018). Social robots for education: A review. *Science Robotics*, *3*(21), 21. doi:10.1126cirobotics.aat5954 PMID:33141719

Bersin, J. (2020). *The skills of the future are now clear: And despite what you think, they're not technical*. Academic Press.

Engestrom, Y. (1987). *Learning by Expanding: An Activity Theoretical Approach to Developmental Research*. Orienta-Konsultit.

Fregnan, E., Ivaldi, S., & Scaratti, G. (2020). HRM 4.0 and New Managerial Competences Profile: The COMAU Case. *Frontiers in Psychology*, *11*, 11. doi:10.3389/fpsyg.2020.578251 PMID:33329228

Lanzara, G. F. (2006). *Capturing Transient Knowledge in Design and Innovation Processes*. The 6th Social Study of ICT workshop (SSIT6) at the LSE.

Pinto, D., Scaratti, G., & Fregnan, E. (2018). *Giovani, Università e Azienda: il nuovo perimetro formativo*. Pearson.

Pinto, D., Scaratti, G., & Fregnan, E. (2021). *Learning Agility 4.0: ecosistemi e trasformazione culturale*. Pearson.

Porter, M., & Kramer, M. R. (2011). Creare valore condiviso. *Harvard Business Review Italia*, *1*(2), 68–84.

Salvadore, G., Rota, E., Corsi, E., & Colombina, G. (2019, September). Industrial Wearable Robots: A HUMANufacturing Approach. In *Mediterranean Conference on Medical and Biological Engineering and Computing* (pp. 1729-1733). Springer.

Staerkle, C., Falomir-Pichastor, J. M., Pereira, A., Berent, J., & Butera, F. (2015). Global value perceptions: The legitimising functions of western representations of democracy. *European Journal of Social Psychology*, *45*(7), 896–906. doi:10.1002/ejsp.2159

Chapter 5
First Conclusions

ABSTRACT

In the chapter are recollected the knowledge structured up to this point: organizations are called to a responsible and generative sharing of value, with the possibility to act on different dimensions: social and organizational citizenship, value for employability, value for digital citizenship, and value for innovation. Creating value at these levels means responding to the needs introduced by the digital transformation, particularly related to the demand for new skills and a new culture of work enabling the 4.0 world. Educational robotics is a privileged means to act in this direction: educational robotic initiatives can be understood as a way of creating shared social value for participants, more specifically related to four areas of value: skills creation, value for new citizenship, empowerment, inspiring and orienteering for working. The Comau e.DO Experience Project also wants to intercept a need to return value to the companies involved and to the territory to which they belong.

INTRODUCTION

The starting point for this research coincides with a reflection on the **value that today's organizations can help generate** internally and externally, through an overview of how the concept of value has evolved. Today, the key to the creation of value seems to lie in the protection and promotion of resources, in their involvement in a process of exchange and sharing governed by moral principles of social justice, sustainable growth, community, ethics, prosperity

DOI: 10.4018/978-1-7998-8653-2.ch005

Table 1. Value-sharing matrix

Value for Social Citizenship	Value for Organizational Citizenship	Value for Employability	Value for Digital Citizenship	Value for Innovation
Being responsible citizens, oriented towards the future and the development of both the territory and the community.	Being members of the "corporate citizenship", sharing its recognized values and feeling.	Developing skills to work in the world of today and of the future.	Being "digital citizens", able to consciously use technologies and to seize the opportunities offered by the Digital Transformation.	Being able to innovate and to develop continuously, with a view on long-term (sustainable) progress.

and genuine responsibility (Porter, Kramer 2011; 2019). An organization, in a generative key, is a place where instrumental and human resources are combined to achieve a shared goal of social development and value creation.

Commons and generativity are the concepts starting from which authors have been able to identify a **value matrix** transversal to the different paradigms and the various theoretical contributions taken into consideration, which refers to multiple levels of citizenship and community life, as shown in Table 1.

However, the context we are living today is characterized by radical technological innovations (Hess, Benlian, 2015) that have triggered profound changes in the way we work, relate to others and live both the world and the communities we belong to (Vial, 2019). Therefore, the question now is: what are today the new spaces to create social value for organizations, touching both organizational dimensions and everyday life issues (Westerman, Bonnet, McAfee, 2014)? To answer this question, it occurs to deep dive the Digital Transformation scenario, the society and companies' evolution, as well as the consequential request for new skills.

FIRST CONCLUSIONS

The Digital Transformation Scenario Requires the Development of New Skills

Digital Transformation is a process where digital technologies create an impetus for organizations to implement responses to gain or maintain their competitive advantage. This scenario has **considerable implications at a social and organizational level**, as it significantly modifies the way we interact with each other, activities and work processes, hierarchical structures and the organizational culture. On the other hand, Digital Transformation presents **great development opportunities for business and for sharing value with the society**, in a relationship of mutual exchange with the external stakeholders of a company.

Within the Digital Transformation scenario, value creation occurs through the development of knowledge and skills that enable people to live and work in a 4.0 world. Authors have focused on those most suitable to adequately manage the scenario of Digital Transformation and company evolution. They have been grouped into two main categories: STEM Skills and Capabilities. These skills are crucial not only for an adequate management of the innovations implied by the Digital Transformation, but also and above all to anticipate the demands of tomorrow's work.

Organizations can create shared value in a generative and community-oriented key **developing such skills and capabilities by means of educational programs** that place new technologies at the center.

Educational Robotics and Educational Represent a Great Opportunity to Create Social Shared Value

The introduction of robotics and robots in educational context supports modern constructivist and constructionist pedagogical theories by placing **the student at the center of the learning process**, that generates knowledge through experience in a hands-on and situated way. The use of Social Robots in the learning process allows to **enrich the learning experience of social interaction** and offers wide opportunities for a **personalized learning process**. It also promotes **motivation and interest** and allows to **familiarize with robots** in the Fourth Industrial Revolution world, that pays more and more attention to the human-robot interaction.

The main challenges in this field concern technical challenges related to the construction of an increasingly fluent human-machine interaction, through the building of an artificial social interaction closer and closer to the human one characterized by a wide range of cognitive mechanisms (Belpaeme, Kennedy, Ramachandran, Scassellati, Tanaka, 2018) and the involvement of all stakeholders in the process, so as to consider their concerns and perceptions, as well as to make Robotic Education an effective and widespread practice.

Educational Robotics fosters a learning environment that is **experimental, collaborative and based on problem solving and critical thinking processes**. It makes students more interested and engaged in **technical subjects related to Science, Technology, Engineering and Mathematics (STEM)** leading to better results and to the **development of transversal skills such as problem solving, critical thinking, team working and collaboration**, defined as the skills required for the 21st century.

One of the main challenges for the implementation concerns the training of teachers and their engagement in these new forms of teaching: **the teacher plays the role of facilitator and can create the right learning context** that allows Educational Robotics to express its full potential. Another major challenge has to do with Educational Robotics diffusion. Within schools, **Educational Robotics should not be limited to after-school courses but should become part of the students' curricula** (Alimisis, 2013). Furthermore, it has the potential to be **applied in more contexts** than just the school one. The introduction of Social Robots and Educational Robotics will thus be able to **support, from an educational point of view, the changes introduced by the Fourth Industrial Revolution** forming the 4.0 Workforce.

Desk-based research has allowed to identify the positive elements of Educational Robotics and, more broadly, of educational initiatives involving the use of a robot. The authors position is that **such Educational Robotic initiatives can be understood as a way of creating shared social value for Participants**, more specifically related to four areas of value:

- Skill Creation
- Value for New Citizenship
- Empowerment
- Inspiration and Orienteering for Working

The Comau *e.DO Experience* Allows to Identify All the Levels of Value Creation Generated by Educational Robotics Initiatives

Analyzing in detail the key elements that characterize the innovative educational robotics project leaded by the Comau Academy - *e.DO Experience* - the first result obtained is the definition of the **main benefits for the Participants**: both teachers and students. These benefits seem to coincide with the areas of value identified through the previous desk-based research activities.

In terms of value creation, in fact, Educational Robotics initiatives such as the *e.DO Experience* seems to serve several functions, particularly related to their **effects on the Participants**:

- Skill Creation
- Value for New Citizenship
- Empowerment
- Inspiration & Orienteering for Working

The evidences suggest that the *e.DO Experience* brings to light **two additional levels of value creation**, related to the Educational Robotics initiatives. In fact, this kind of initiatives seem also to direct the creation of value at a **company and territorial level** in terms of commons and generativity.

More in detail, the **Value for Company** seems to be related to the following areas of value:

- Core Business/Core Mission
- Brand (Reputation/Marketing)
- Employee Branding
- Customer Social Responsibility (formal obligations)
- Customer Social Responsibility (true commitment to create a better world)
- Open Innovation (get the innovation potential that is present outside the organization - talents, know how, ideas)
- Employees' Exposure

The **Value for Territory**, instead, seems to be related to the following areas of value:

- Creation and Dissemination of Culture
- Didactic Innovation
- Contribution to the Development of the Skills required from the World of Work

Figure 1. The summary synoptic

LEVELS OF VALUE	AREAS OF VALUE & INDICATORS
VALUE FOR PARTICIPANTS	SKILL CREATION
	o Robotics Skills
	o School Subject Skills
	o Professional Hard Skills
	o Soft Skills (communication, problem solving, negotiation, etc.)
	o Working in the Future Skills (Capabilities)
	▪ Deal with Human
	▪ Deal with Technology
	▪ Agility
	▪ Engagement
	▪ Collaboration
	▪ Interdisciplinarity
	▪ Innovation
	VALUES FOR NEW CITIZENSHIP
	o Challenging and Stepping out of Comfort Zone
	o Respecting Others
	o Being Exemplary and Consistent
	o Keeping Promises
	o Measuring Results Objectively
	o Cultivating Responsibility and Reciprocity
	o Pursuing Equity and Justice
	o Being Transparent and Sincere
	o Being Open to Comparison and Exchange
	o Meeting the Timeframe
	o Celebrating Success
	EMPOWERMENT
	o Being Satisfied and Enthusiastic for Overcoming a Challenge
	o Being Aware of your own Qualities
	o Being Confident and Constructive
	o Being Passionate and Energetic
	o Being Proactive and Enterprising
	INSPIRATION & ORIENTEERING FOR WORKING
	o Having Galvanizing Experiences
	o Getting in Touch with Professional Models
	o Providing Professional Trajectories
	o Developing Greater World of Work Awareness
	o Developing a Greater Awareness of One's Strengths and Weaknesses
VALUE FOR COMPANY	CORE BUSINESS/CORE MISSION
	BRAND (REPUTATION/MARKETING)
	EMPLOYEE BRANDING
	CUSTOMER SOCIAL RESPONSIBILITY (FORMAL OBLIGATIONS)
	CUSTOMER SOCIAL RESPONSIBILITY (TRUE COMMITMENT TO CREATE A BETTER WORLD)
	OPEN INNOVATION
	EMPLOYEES' EXPOSURE
VALUE FOR TERRITORY	CREATION AND DISSEMINATION OF CULTURE
	EDUCATIONAL INNOVATION
	CONTRIBUTION TO THE CREATION OF SKILLS REQUIRED BY THE WORLD OF WORK
	CONTRIBUTION TO THE CREATION OF NEW ECOSYSTEMS

- Contribution to the Creation of New Ecosystems

CONCLUSION

The evidence gathered made it possible to identify the **specific values of Educational Robotics initiatives** in terms of commons and generativity. Authors processed the results that emerged to **map all the elements of value** generated by Educational Robotics initiatives, having as a reference the *e.DO Experience* project.

Here you can look at the summary synoptic, which shows:

- **levels** of value (Participants, Company, Territory)
- **areas** of value for each level
- value **indicators** referring to each area.

The next step of the research work is the **verification of this synoptic**, as expressed by research goals:

1. Verifying and exploring the values generated by the *Comau e.DO Experience* Educational Robotics Project and identifying, listing and describing the main indicators for each value;
2. Measuring the weight of each value and its respective indicators in the perception of those who designed and delivered the *Comau e.DO Experience* Project.

To achieve such goals, two different tools have been set up:

- An **interview outline** to be administered to a sample of Comau managers and Comau partners involved in the design and implementation of the *e.DO Experience*, to verify and investigate the elements of the synoptic by means of the analysis of their perception concerning the value of this project;

- A **survey** aimed at a larger sample of Comau managers and Comau partners to measure the weight of the elements reported in the synoptic.

A variety of different perspectives has been captured, so as to **portray a complete and in-depth picture** of the different levels of value generated for the participants, the companies and the territory involved, taking into consideration the **point of view of Comau managers and partners** who designed and delivered the *e.DO Experience* project. The next chapters will describe the results of these research activities.

REFERENCES

Alimisis, D. (2013). Educational robotics: Open questions and new challenges. *Themes in Science and Technology Education*, *6*(1), 63–71.

Matt, C., Hess, T., & Benlian, A. (2015). Digital transformation strategies. *Business & Information Systems Engineering*, *57*(5), 339–343. doi:10.100712599-015-0401-5

Porter, M., & Kramer, M. R. (2011). Creare valore condiviso. *Harvard Business Review Italia*, *1*(2), 68–84.

Porter, M. E., & Kramer, M. R. (2019). Creating shared value. In *Managing sustainable business* (pp. 323–346). Springer. doi:10.1007/978-94-024-1144-7_16

Vial, G. (2019). Understanding digital transformation: A review and a research agenda. *The Journal of Strategic Information Systems*, *28*(2), 118–144. doi:10.1016/j.jsis.2019.01.003

Westerman, G., Bonnet, D., & McAfee, A. (2014). *Leading digital: Turning technology into business transformation*. Harvard Business Press.

Chapter 6
Educational Robotics for Value Creation:
Qualitative Interviews

ABSTRACT

Through a qualitative survey that involved the use of a semi-structured interview, the authors wanted to investigate, within a sample of Comau managers and partners, which elements of value for participants, company, and territory had been generated by the Comau e.DO Experience Project. The objective was to describe the experiences, images, and representations associated with the value produced by the e.DO Experience. The value for participants was explained in terms of skill creation, value for new citizenship, empowerment, and inspiring and orienteering for working. The value for company was described not only in relation to the development of the brand or the core business but also and above all in function of an authentic social responsibility and growth/recognition of internal staff. The value for territory concerns instead the creation of ecosystems between company, community, institutions and territory, educational innovation, cultural dissemination, and promotion of the development of capabilities.

DOI: 10.4018/978-1-7998-8653-2.ch006

INTRODUCTION

The starting hypothesis is that the Educational Robotics initiatives undertaken by Comau through e.DO Robot and the *e.DO Experience* Project have enabled the **generation of shared value among Participants, Company and Territory**. The assumption is that the many promoters, whether they are Comau managers or partners of third-party companies, have been driven by shared goals and interests but, at the same time, their perceptions and representations of the different activities and the value they have generated may present peculiar and unique elements. Here the goal is to **capture the variety of different perspectives**, so as to portray a complete and in-depth picture of the **different levels of value generated for the territory, the companies involved and the participants**.

The research objective involves describing a series of value-sharing experiences implemented through the e.DO robot. Consistently with the purposes mentioned above, the methodology employed is that of interpretive phenomenology, whose main object of study is the understanding of an experience (Richards, Morse, 2008). This qualitative phase is designed to meet the needs of exploration of the phenomenon studied; in fact, it is intended **to provide preliminary information for understanding the topic studied and to direct the creation of an in-depth survey**, discussed in the next chapter.

This phase of the research involved the **administration of 14 interviews**, one of the privileged tools for collecting data within the phenomenological approach, **equally divided between Comau managers (7 interviews) and Comau partners (7 interviews)**. This choice of sampling made it possible to involve representative actors who were heterogeneous in terms of role, vision and - in the case of the partners – the organizations they belonged to. Their contribution made it possible to outline elements in common or divergence between the various organizations involved, so as to provide indications for structuring the subsequent survey, which will be addressed to a larger sample.

Figure 1 shows the interview outline.

Figure 1. Interview outline

1. Name and Role
2. Short description of the educational robotics initiative (name, participants' description, place, time, goals and program, results)
3. What did your company/organization and you deal with within the educational robotics initiative?

SECTION A - OVERALL

1. What value do you think the proposed initiative has for the PARTICIPANTS (what was it for, what did they bring home)? Tell an episode to exemplify the value you perceive.
2. What value do you think the proposed initiative has for the COMPANY/ORGANIZATION (what was it for, what did they bring home)? Tell an episode to exemplify the value you perceive.
3. What value do you think the proposed initiative has for the TERRITORY (e.g., scientific community, municipality, citizenship, institutions, families, etc.) (what was it for, what did they bring home)? Tell an episode to exemplify the value you perceive.
4. What is the image (concept) that best represents the value / values you have identified?

SECTION B – AREAS OF VALUE

1. Compared to your initiative, how important have these areas of value been during its design and implementation? *Considering 100 as the total, give a percentage weight to each of the three areas.*

- VALUE FOR PARTICIPANTS:
- VALUE FOR YOUR COMPANY/ORGANIZATION:
- VALUE FOR THE TERRITORY (e.g., scientific community, municipality, citizenship, institutions, families, etc.):

2. Do you identify other areas of value? If so, which ones?

SECTION C – PILLARS OF VALUE

Below is a list of values, divided in three areas of value. Among the proposed ones, select the values that you believe are generated by your initiative.
For those selected, explain why.

1. Values for participants

AREA of VALUE	VALUE	YES	NO	WHY (for those selected)
PARTICIPANTS	Skills Creation			
	Values for New Citizenship (Human Values)			
	Empowerment			
	Inspiring & Orienteering for Working			

2. Do you identify other values? If so, which ones?
3. Values for your Company/Organization

AREA of VALUE	VALUE	YES	NO	WHY (for those selected)
COMPANY / ORGANIZATION	Core Business/Core Mission			
	Brand (Reputation/Marketing)			
	Employee Branding			
	Customer Social Responsibility (formal obligations)			
	Customer Social Responsibility (true commitment to create a better world)			
	Get external innovation potential (talents, knowhow, ideas)			
	Employees' Exposure			

4. Do you identify other values? If so, which ones?
5. Values for the territory (e.g., scientific community, municipality, citizenship, institutions, families, etc.)

AREA of VALUE	VALUE	YES	NO	WHY (for those selected)
TERRITORY (e.g., scientific community, municipality, citizenship, institutions, families, etc.)	Creation and dissemination of culture			
	Didactic Innovation			
	Contribution to the creation of new ecosystems			

6. Do you identify other values? If so, which ones?
7. If you have identified another area of value, try to indicate the corresponding values.

DATA COLLECTION AND PROCESSING

An overview of the interviews is presented in this section

- Respondents' characteristics, by reporting their answers to the first 3 framing questions.
- Answers related to SECTION A - OVERALL organized by value levels (Participants, Company and Territory);
- Analysis of the images described by the respondents;
- Respondents' answers referring to section B – AREAS OF VALUE;
- Summary synopsis of the main elements collected during the interviews through the questions contained in Section C-PILLARS OF VALUE.

Qualitative Interviews: Comau Managers

Interview 1 – Marta Serrano

Name and Role

Let me introduce myself, I am Marta Serrano and I have been a consultant for Comau for almost three years in the Academy area, which is now called the Education & Consultancy segment. I was called to Comau to deal with Business Development, that is the whole area of business development of the e.DO Experience program in Italy.

Short description of the educational robotics initiative (name, participants' description, place, time, goals and program, results)

As e.DO Experience, we have worked on three specific products in particular: e.DO Learning Center, e.DO Learning Lab and e.DO Licensing. The Learning Center revolves around the figure of the e.DO robot, which is a six-axis robotic arm that wants to emulate what are the functions of an industrial robotic arm; it has been completely adapted, especially in terms of safety, but also in terms of design and ease of maneuvering to be used by the younger children, even from eight years upwards through just one tablet. All of us, I believe, are now fortunate enough to be in possession of a tablet and, with that single device, many activities can be carried out at educational level.

What did your company/organization and you deal with within the educational robotics initiative?

Certainly, this need to bring the generations closer, from an early age, our children, bring them closer to everything that is the world of technology and especially the world of STEM. Such school subjects are today still perceived at school level as the ones that you do not necessarily have to like but, through the use of this robot, we wanted to show young people that it is not necessary to like a school subject, rather you have to understand the applicability, why to put yourself in front of, for example, the associative property and the commutative property that a teacher explains to you through a robot you can really understand how this will be applied to your future life, in a future professional profile that you have to start to imagine even if it is true, it is now written everywhere, that 70% of the jobs that our children will do in a few years still do not exist, we travel at a great speed but, in short, the basis of technological and scientific level remain.

Interview 2 – Nicoletta Beretta

Name and Role

My name is Nicoletta Baretta, I arrived in Comau at the end of 2018 with a specific goal: to develop projects and the business of education through technologies and robotics in markets other than Italy, that is, in EMEA markets, and then they have also become some extra EMEA ones, because I also follow the United States, India, LATAM with some colleagues. Then I work together with the Academy team, that is then Academy, today Consultancy & Education, because today is a real business segment.

Short description of the educational robotics initiative (name, participants' description, place, time, goals and program, results)

And so, we started our collaboration activity with Comau and in the team, when I arrived, there was already a very interesting project that was called *e.DO Experience* that involved the use of a robot with an industrial concept but then adapted to education and that was addressed, but still addressed even if certainly giant steps have been made compared to the end of 2018.

In schools it is highly appreciated that a robotic arm, with a gripper and a marker holder, lands on the tables of children and teachers, a set of tools such as precisely a Cartesian plane, the soul of the robot to be assembled at the beginning of the activities. Here comes this change of approach to methodology, this change of approach to learning that creates a great educational experience, it creates engagement, motivation, but not only for the students, also for the teachers who find a new way, a new method as well as a new tool to, how to say, wake up the minds of their pupils, their students at all levels, because we start with students from 7-8 in primary school that were the main target, how to say, to which we addressed; then the project led us to work at different levels, presenting projects to universities, training institutions, professional institutes.

What did your company/organization and you deal with within the educational robotics initiative?

The challenge is to import an industrial robot into education. The world is changing very fast because of the Digital Transformation, the big challenge is to prepare young people for the upcoming world, to prepare them for Industry 4.0.

Interview 3 – Giuseppe Daqua

Name and Role

I have been working with Comau since 2016, just to frame the time a little bit, I started with a collaboration as an external consultant and then from 2018 I was also hired in Comau. I started my journey with Comau to follow the e.DO Robot project. I have a scientific background, I have a degree in physics and Comau was looking in 2016 for a figure to follow scientific training projects, so with robotics, through a new tool that did not exist yet. What do I deal with in Comau? I am part of the team that we call the Academy team, but I'm part of the team that designs training paths for e.DO robots, so the first part is design, I train delivery, then I go into delivery, I lead the courses and then I train customers or people who need to be trained, so if I have to give you three keywords for my activities in this area of e.DO, they are the design of paths, delivery and training.

So, these are three macro areas that I deal with: development, delivery and training.

Short description of the educational robotics initiative (name, participants' description, place, time, goals and program, results)

We deliver classes with e.DO but not robotics ones, this is a very important issue for e.DO, lessons with robotics. Robotics for us is not the purpose of what we do with e.DO, it is the tool, that is, we start from e.DO as a tool, as an interesting tool, and then we convey other contents that are, if we talk about the school world, the curricular subjects. So, in what is our main program with e.DO, the Learning Center, I give math lessons with e.DO, I give computer science lessons with e.DO. That was the idea of the first project. And in fact, a world of initiatives revolves around e.DO. We have also included it in more professionalizing paths and consequently of introduction to the world of work, such as robotics license or robotic license. Comau certifies that the student knows how to use and program an industrial robot and it is a certification that you put in the curriculum and you spend in the world of work. In the resume you can write that a company like Comau certifies that you have completed a course of use and programming.

What did your company/organization and you deal with within the educational robotics initiative?

e.DO it's a tool that goes a long way, that is, it was a great challenge for Comau because it is not an industrial robot, I want to say that right away. Comau is a company that builds industrial robots, and therefore for heavy industry, for industrial work, e.DO is a robot that has the logic of an industrial robot, which was built by a company with that experience, 50 years in robotics, but, on balance and also at the level of legislation, it is not an industrial robot, it is an educational robot. It is a robot that Comau has developed precisely to bring anyone, and I really mean anyone, closer to the world of robotics, of new technologies but also of mathematics and therefore of the subjects that are studied at school. Now I am focusing on the student's side of things, but the e.DO project is a bit broader than that.

Interview 4 – Stefano Pesce

Name and Role

I have been in Comau for three years now, I started my career in Product Development in robotics and for the first year I was in charge of the software and software architecture within the e.DO educational robot; after that I became Product Manager, worldwide, of the robot and of the *e.DO Experience* in general.

Short description of the educational robotics initiative (name, participants' description, place, time, goals and program, results)

With *e.DO Experience* we mean everything that has to do with e.DO, and therefore also all the educational packages, the global solutions that we propose with e.DO. Essentially, e.DO is an educational robot of small size that can carry 1 kg on its dashboard, the goal of this robot is to try to teach, to the largest possible number of people, robotics but not only; we also use the product for educational initiatives for STEM subjects and then large scale on mathematics, coding and many other subjects covered and then the robot becomes essentially a tool, a means to get to the students thanks to the teacher, who is always a central figure. What we have created during the years is this educational tool, we have mentioned the teacher, but it engages especially the students of all age groups, because our target audience is really wide: it goes from young children up to even, we did courses at the university of the third age, 90 years. When you turn it on and you see the movement of a robot that is not a toy but essentially an industrial robot within the educational metaphor, at that moment it becomes an educational object itself, and it is also able to attract both students and teachers, so this is certainly the main peculiarity of the object.

What did your company/organization and you deal with within the educational robotics initiative?

What we did as Comau was, unlike many other big competitors, that we started from the industry, Comau is a big robotics company with a huge knowledge of the world of robotics and the automotive industry, we started from these concepts and brought them into the educational world. Now many, without doing so much benchmarking, but for example Lego or other tools start from the world of toys and bring, within the world of toys, robotics, because obviously it is an important trend. And the path is a bit different: on the one hand, there are companies that start from the bottom and bring the knowledge above their level; we started from the top of robotics and we have dropped the robotics tool within the educational band of the early years until, precisely, the university and then going forward in the training of employees. When a company is proposed as educational robotics but starts from a strong educational level, it is easier to enter the world of schools. We immediately experienced, let's say, not a stumbling block, but a bit of friction because we presented ourselves as the great Comau, expert in robotics, in creating machines and cars, and we then dropped our concepts into the world of education. But there was an incredible success at the beginning, precisely because of this: because our strong value proposition was to bring the world of industry, the professional world, right from the start into the school and therefore, even today, we are configured as a link, the connection between school and business, and there in the middle we can really have our say.

Interview 5 - Francesco Faccio

Name and Role

I work with Comau, I am an external contractor outsider, I collaborate with Comau practically full time and I have been collaborating for two years with the Academy, Comau Academy, for the development of upskilling and reskilling, so let's say development and change, integration. Therefore, I have collaborated, I still collaborate, with Comau on this issue and we have worked a lot together to really focus the concepts and the value of what can give this use of technology in the educational field, or better, the development of technologies dedicated to education.

Short description of the educational robotics initiative (name, participants' description, place, time, goals and program, results)

Coming back to us, we have developed some paths using in particular e.DO, but also the exoskeleton, to pass concepts. We worked for the kids, we built the Learning Center, but my colleagues will have already told you about it, where students from 8 to 19 years old, having fun with technology, have learned and are learning to draw an ellipse, Boolean logic, with programs using robots. It is amazing how much easier it is, they are accepted spontaneously, without problems, by 8 years old kids rather than by 40 years old executives.

What did your company/organization and you deal with within the educational robotics initiative?

We come from the industrial world, not the educational world. In my opinion we were the first to make this leap and our competitors are making the same leap now, and then there are companies that already made robots for gaming or for Brico that are also coming to education, but they do not come from the industrial environment, so this is not very common. In the Pulp Fiction movie, there is a scene where John Travolta and Samuel L. Jackson accidentally shoot a guy in a car, then they go into a crisis, they get to the house of the friend shocked, they are in full panic and then they call the boss, who sends Wolf. Then Wolf rings the doorbell and says "Are you Bernie?" "Yes, I'm Bernie, who are you?" "I'm Wolf, I solve problems". Well, we solve problems, that is what we do. What kind of problems? Well, the problems are making people and processes work together, in an unstructured way, in an environment that is changing so much due to the digital revolution. Beware that this stuff that has revolutionized our lives, maybe not the digital natives, my son is 17 and for him Google is the most natural thing in the world as all the other things here, but we have revolutionized it a bit and many others like me have adapted their life, but just daily life, and then we go to the company and send the fax or wait for the fax. This thing that has revolutionized your life, you have to realize that it is already revolutionizing you, and even more the way you work. If you are a manager and you do not adapt, if you are a worker and you do not adapt, if you are a company and you do not adapt, in 5 years you will be out of the game because Digital Transformation has an exponential evolution.

Interview 6 - Lucia Marchesi

Name and Role

Since 2017 I have been in charge for Comau of the development of all the training packages of the *e.DO Experience*, I manage the development team and all the Customer Operations, so I have a more management and managerial role, and I am also responsible for all the creative part, the development, the conception of everything that is around the training of the Comau Academy to the outside, and it all started from the *e.DO Experience*, so we started this way with the creation of the first training packages that we delivered to the Fondazione Agnelli.

Short description of the educational robotics initiative (name, participants' description, place, time, goals and program, results)

Everything started with the first training packages with Fondazione Agnelli, then from there we grew exponentially in the last four years, in an incredible way.

We do training with e.DO, as curricular content what we convey is mathematics for primary and secondary school, and then programming for primary and secondary school, university and adults, so the different programming languages, but especially what we define as curricular, so programming, which should be the minimum. Then we do a lot of simulations of business processes, so Factory Simulation, from the secondary school up we work on the difference between effectiveness and efficiency within an organization, the distinction of roles within the organization, how you distribute them, with the difference between production in series and in parallel, then working on mindset that is already organizational. It also applies to the average, are light years away from maybe start working, but start thinking in these terms can also help them in the organization of their daily lives, of their small projects.

What did your company/organization and you deal with within the educational robotics initiative?

The main challenge was precisely that of setting up the educational model. The facilitator is the one who facilitates the learning process of the people in front of him or her, and therefore who works on the needs that emerge, the one who manages the dynamics of the classroom and therefore is able to work precisely on the basis of the feedback that can be detected to maximize the learning results. We have called this role facilitator because the training model we propose in the e.DO Experience is that of the "train the trainer", so we train the facilitators who then manage these workshops and when we do the "train the trainer", apart from all aspects of the importance of the classroom relationship on which we work so much, on the aspects of feedback, on the aspects of teamwork and so on, we work on the concept of how to read the learning needs of people in training, then all aspects of management, confrontation, exchange and interaction and how it is important, with what phrases, in what way, to facilitate the learning of all the participants.

Interview 7 - Malcolm Bisio

Name and Role

Now I am in charge of the commercial aspects of the educational part, so not only the strictly hardware part, e.DO and the various components, but also the sale of the whole solution. I have been dealing with this part since July/ September 2020, so it is a relatively short time. Before doing this job, I was dealing with delivery and training, so I was part of the Academy and especially in charge of the delivery part. I delivered a lot of initiatives in Fondazione Agnelli, then in the Learning Center that we used before Covid in Fondazione Agnelli, training to trainers, to Comau customers who were going to buy our solutions, events and everything. During the more "Academic" part, let's say, concerning delivery, obviously there was also the whole product development phase, so the development of training packages. That is a part that I touched for a quite short time, let's say, before moving then to the commercial area.

Short description of the educational robotics initiative (name, participants' description, place, time, goals and program, results)

The *e.DO Experience* offer is composed of some packages, I will focus on what I think is more central for this type of research. However, products that concern foundations, municipalities and therefore entire solutions provide the purchase of a certain number of materials, training packages, training to trainers and, in this way, the institution on duty is able to provide the territory with a service, typically a free service to schools participating in these initiatives. Another fundamental part of which precisely I also knew a little more product development, is that which concerns instead educational packages but for schools, then solutions a little smaller, both from the physical economic point of view, even the educational packages are a little shorter, and you go directly to propose it to schools, so that the school does not have to participate in something larger outside, or rather if you want you can do it internally, perhaps with a small laboratory of robotics or computer science a little cutting edge with robots. So, there is a service that is provided directly for schools and one that goes through foundations, municipalities, Chamber of Commerce, in short, institutions in general. Now the *e.DO Experience* is much bigger than that, and it also refers to the whole professionalizing part, it starts and it crosses with the world of schools, through technical institutes that may have needs both from the classical STEM and professionalizing points of view, and then from there it connects to universities and vocational training centers. Then you start from there and you deliver a more technical training, that is more specific, even a little bit longer, collaborating with employment agencies, with different training agencies, in short, the world and the ecosystem expands a bit more.

What did your company/organization and you deal with within the educational robotics initiative?

You learn to bridge, or rather to not bridge, the generation gap, because you cannot, but to translate things from one way to another. It is a way of communicating that I think is also good for the company, because then you find the same people in the company, you find people who are your age, with whom you do not even need to talk so many times, and you find people who are three times your age and have difficulties in speaking the same language as you, so experience in the field certainly makes it easier for you when you return to the company to communicate as well as to do it in a more functional and effective way.

Qualitative Interviews: Comau Partners

Interview 1 – Paolo Cesana (Fondazione Clerici)

Name and Role

At Fondazione Clerici, we are all part of a Confederation of which I am the National Secretary, the Confederation of Christian Matrix Institutions, which is inspired by the values of the Social Doctrine of the Church, where the dimension of inclusion is central and inclusion is achieved through the attention to people that starts from the principle of educational care, i.e., following the children through their weaknesses, understanding what their weaknesses are, building customized paths to ensure that they can find full self-realization.

Short description of the educational robotics initiative (name, participants' description, place, time, goals and program, results)

For us, e.DO is part of a project aimed at teasing out the inclinations of children and make sure to develop, to leverage the motivational dimension in traditional education, so that in the classroom the children, using a robot, can experience active experiences.

"skillstrainingspace" is a project that we created aimed basically to talk to the heart of the boys to bring out the inclination of younger children, children of middle school, primary school, but then continues for elder students at high school to ensure that they can over time, through robotics, coding and technologies in general, experiment and, through an approach let's say more playful, train a little their own baggage of internal skills, pulling them out and then making the right choice, or more than right, a more enlightened one, more oriented than what is their own path of study, their career studies.

What did your company/organization and you deal with within the educational robotics initiative?

The need is to touch the paradigms and the culture of education, its methodologies. So, in essence, we have been working for years through a method, a methodology that is inclusive and active for children starting from the laboratory dimension, we do both school and vocational training borrowing a lot of experience from vocational training in school. The goal we have set ourselves is to say "let's make sure that, especially on some subjects where it is more complicated, especially at school it happens that for subjects such as mathematics, physics, we apply more or less the same methodology that we apply in the professional area" and then basically through the laboratories the kids have fun and meanwhile they learn STEM, math. We say that teachers go on like this, it is a machine, an industry of the enlightened type, we have this type of social setting for which when a child, and there are now more and more, has fragilities such as ADHD, he or she tends not to be aligned with the traditional system with the risk of being "cut out", discarded and therefore, the fact of discarding them becomes then a social problem. Our system, that of vocational training, which starts from the approach of doing, the apprenticeship and the workshop approach and from the idea that young people have intelligence in their hands, young people put themselves to work and then, through the lever of motivation, little by little they come to develop knowledge. Therefore, the lever is motivational, that is to say, a kid who maybe is compulsive oppositional says "I don't want to do that stuff, I don't do it because I'm not interested, not because I'm not capable" or someone who has ADHD arrives home tired from school because the bird flying outside the window attracts the same attention of the teacher, right? So, in the end one is dazed and the other comes home with the head aching, you have to work on the motivational dimension with these kids.

Interview 2 – Manuel Tonolini (Fondazione Dalmine)

Name and Role

I am Manuel Tonolini and I am in charge of the education sector at the Dalmine Foundation. I started working on education about ten years ago when the Dalmine Foundation, which was created mainly as the historical archives of Tenaris Dalmine, developing a specific project for schools. So, until ten years ago, we worked with schools on an occasional basis when we did exhibitions and so on, and then we said, "Ok, since the schools are asking us for...", we threw ourselves into the field of education, as a response to the needs expressed from outside.

Short description of the educational robotics initiative (name, participants' description, place, time, goals and program, results)

We bought five e.DO, at the beginning there were three educational packages and then they became nine when Comau expanded the offer, we made an agreement with BergamoScienza, which is the association that deals with this Festival of Science that is now more than twenty years old. We did not train internally, but we relied on those who already worked on these issues in the area and five educators of BergamoScienza went to Turin and did all the training, then began to deliver the Comau courses, so those organized with them on our territory and this thing lasted two school years: one school year finished the training, it finished in January, it was 2019 so all of 2019 and part of 2020.

It was the school that came to us, because otherwise it would have been too complex, if not for some of the programs, let's say collateral, so sometimes we brought the robots for a week in a school and there we did a series of workshops, but on robotics basically they were the ones who came to us with a schedule, exactly what basically happens at the Agnelli Foundation, and they had their activity.

As robotics and Comau packages, looking at the numbers, let's say until February 2020, we basically trained half of the kids from primary and half from secondary school. They had a great interest until the secondary school, especially high schools or non-technical institutes in the sense that the more technical part for us was more discovered, but also because with e.DO we would have had to do a whole other series of jobs that we never did, the purchase of a whole series of things that would have allowed us to make an extra leap on the professionalizing side that we had not yet done. As educational goals there it was mainly let's just say STEM in a general way so they actually, it was kind of what I was saying at the beginning, explored a little bit the world of math, geometry, Cartesian axes, etcetera using the robot and a little bit of coding which then increased. So basically, if I look at it in general, basically math, geometry and a big increase in coding in the last few years; third thing group work was another key element that e.DO pushed so much and that schools had started to propose and on which we saw big differences: there were schools that were already accustomed to it, so you could see the small group that had dynamics, structured logic because they were already used to it, and those that were "one hammer and the other screwdriver", in short, were not used to working in groups.

What did your company/organization and you deal with within the educational robotics initiative?

Ten years ago, because this experience started ten years ago, there were some teachers, mostly from secondary school, who began to understand that the didactic activity as it had been done in the last 150 years was beginning to have its cracks and that they had a historical archival heritage in the area, a nice place, because the Foundation is really in a very nice place, where they could do didactic activities and there was a staff that responded, in the sense that we were there and we kindly offered what we could. With them we began to think that it might be interesting for a Foundation like ours to enter the world of schools, and that is where the theme began.

Then the relationship developed in a bidirectional way because the schools clearly started to ask us for things and we started to offer them, we always seized the ball, we want to say that then, once we understood the system and where it was going, we would make proposals and then every year we would make our proposed workshop or projects in the area and actually build them with this group of teachers who then gradually created in the territory.

Interview 3 – Matteo Pisanu (Digital Innovation Hub)

Name and Role

My role is Head of the Digital Innovation Hub Vicenza, which is not a spin off, but we could define it as a Business Unit of Confartigianato Imprese. It was born in 2016, it is accredited at European level and it is part of the Industry 4.0 Network, as it was imagined at the time by the former minister Calenda; it currently has a workforce of 15 people who are engaged, precisely, in accompanying small and medium-sized enterprises, in particular small artisan enterprises and not, of the Vicenza area in the path of digital transformation or digital transition, as we use to say recently. In the last two years, we had an explosion of requests and we also made ourselves known, if before we had to be the ones to, how can I put it, stimulate an interest on the part of businesses, now it is easier for businesses to contact us, so you can see that there has definitely been a change of pace.

Short description of the educational robotics initiative (name, participants' description, place, time, goals and program, results)

Through a project co-funded by the Cariverona Foundation, we worked together with Comau for the establishment of a certified Learning Center, so it was two years ago that we purchased the Comau technologies for the Learning Center, which consist of four robotic arms and e.DO and in four Desk Simulators, this allowed us and still allows us not only to provide training packages related to educational robotics, but also to work in another segment that is that of industrial robotics through Desk Simulators, so let's say we can cover, through the Learning Center, two strands that are, on the one hand, the pedagogical one, more related to the school world and, on the other hand, the one that was born as a robotics license, but that we can offer to employees of small businesses as a qualification, let's say, to the use of industrial robots. In the first case, the use of the robots is linked more to the teaching of curricular subjects at a transversal level with all the aspects of soft skills that emerge, while, on the other hand, it is already something more directed towards the development of work skills.

What did your company/organization and you deal with within the educational robotics initiative?

Well, putting aside the issue of the pandemic that has helped on some fronts when it comes to digital, helping companies to have an acceleration that maybe we would have had in three years and we actually had in three months, apart from this, I think that the change of workforce within companies has also facilitated an awareness or probably even economic incentives have been a lever to bring some models, some systems within companies. Maybe some things were not thought so because often companies, especially small companies, buy a machine also for the sake of having a new machine, right? And maybe they exploit just 50% of its potential, then maybe who manages all the infrastructure, slowly inserts elements of digitization, technological advancement. I think that this is the result of a process that was inevitable and that could have been much faster than it actually was, for example it is enough looking to other countries, so it is not that there have been great developments, but there has been an improvement. There have been two main challenges: from a certain point of view, there has been an effort by the Comau team of developers to develop a technology that would be a link between the world of educational robotics, as we have always known it, and industrial robotics. That is, although it is a robot, how is it called, personal care robot, that is a robot that does not create any kind of problem to users, on the other hand, in short, the degrees of freedom, the operating characteristics, although simplified, actually go to simulate the industrial robot perfectly. There are many things that with the technologies we used before were not possible, so it opens a spectrum of possibilities that we did not have before.

Interview 4 – Matteo Bonanno (Pearson Italia)

Name and Role

I work for Pearson's Academies division, everything related to teacher training. In particular my Job Title, so what I deal with specifically, are all projects in the STEM area with a focus on robotics. I have followed the project from the beginning, since Pearson and Comau have made an agreement for the development and release of the products on the market, then I gradually became a reference on the technical side from Pearson, so I am the interface for the design aspects with Comau. We do training for school teachers from all points of view, so about teaching methodologies, for what concerns the disciplinary aspects, or the whole disciplinary theme and the technical resources with which they teach.

Short description of the educational robotics initiative (name, participants' description, place, time, goals and program, results)

The projects with Comau are linked to training and certification, particularly aimed at students. They are initiatives that bring the use of professional applications, using a blended mode of training that provides contact with the teacher and the possibility for people to train themselves; everything has been structured so that it has a flipped approach, a cooperative approach within the laboratory activities, and then you come out with a certified competence. I think our big intuition has been that an industrial robot can become an educational robot, so e.DO has that value, the same object can accompany you from the age of eight up to your profession, the same object used in different ways. How do our projects work? Our partnership with Comau is very broad and the project has been co-built from the beginning, it is not that Pearson just latched onto the initiatives that Comau asked for. We sat down at the table and said "Ok, there is this Pilot, how do we make it scalable across Italy?" and we came up with a model together. What kind of model? Basically, Comau provided its pro platform, then Pearson integrated the contents of the platform by building and adapting it to the former alternating school-work paths, in addition we included what we call "Module 0" on the world of work and cooperative learning, just as a framework for the famous soft skills in a hard path. And then we took care of all the delivery in the marketplace and all the training. So, the trainers of the project are from Pearson and whoever trains the teachers is Pearson too, whoever does the exams is Pearson, so we basically acquired all the skills that Comau wanted to pass co- building the train the trainer activities. Basically, we took care of every aspect. Comau gave us some tools that we then made our own and used for the diversified initiatives with e.DO., obviously always in partnership, in the sense that there is constant contact because the project is unfeasible one without the other. Comau has the means and the know-how in robotics, we help to act in the dissemination. In any case, we have always been involved, even in some of the steps of product co-building; e.DO had beta tests as part of its development, so we gave suggestions and helped identifying some problems.

What did your company/organization and you deal with within the educational robotics initiative?

The initiative was born out of an experiment that Comau did on its own, so they developed this training platform for professionals and thought they could flood it into other fields. The need that then originated our involvement was also linked to the fact that we have a strong capillarity on the school at national level, so the need had become to expand the project as much as possible and we looked as the most suitable partners. However, we basically shared a need that came from both sides. We are concerned with providing the best tools for future generations, obviously through textbooks because that is our core business, but also through all the related products to be able to face the future. That is the primary focus of our work. They are the people who acquire knowledge from a job standpoint, people who are trained for a long-term job position. We both agreed that there was a lack in the field of robotics: there is a sector in enormous expansion from the point of view of sales, in continuous growth, but also at political level, political in the broadest sense, from European level to the most local level we continue to notice a lack of professionalism in the STEM field in general, but also in the field of automation. On our side, the need is to give the best and most modern tools, the ones that best respond to the context in which young people are immersed when they leave secondary school. The contact was born from there, from the need to respond to a need.

Interview 5 – Diego Teloni (MakerDojo)

Name and Role

I am an entrepreneur. The focus of my entrepreneurial activity concerns two points: one is to build projects of innovation and digital transformation, and let's say it is a classic consulting activity with the attention to build services or products; the other is strongly dedicated to the construction of a series of places or hubs that can promote innovation in the territories. These hubs, whose scope of activity revolves around the enabling role of technology, for instance how technology can foster innovation and inclusive growth, see the theme of Digital Transformation as central, but above all the theme of quality of the human capital, as the more we are able to grow human capital, more territories are able to grow in an inclusive way.

Short description of the educational robotics initiative (name, participants' description, place, time, goals and program, results)

As a network of hubs, we are growing and working on a national scale. MakerDojo, which is the project that does edu-tech, is a national project. We have a consumer segment, for example families and young people, we have an institutional segment that is the set of realities that have objectives of growth of human capital, so the Chamber of Commerce system with the alternation of school and work, cities, local authorities, regions that are responsible for systems of education-training-work, then we also work with some companies for more defined courses. We are partners with Comau, in the sense that Comau has a great capacity for research, development and design. We have the ability to include a certain type of project within an educational offer and a relationship with the territory, so it is a continuous mutual input and feedback and MakerDojo works with Comau on Learning Centers and has a plan for the development of Learning Centers. We are currently working in consumer on User Programming robots' courses, then periodically we discuss new joint initiatives. Comau has done an initial set, they have done some interesting design work on robots. The project includes packages, that are carried out by trainers who must be certified, and this is an element of guarantee and quality for everyone; we have several certified trainers or educators for the different formats developed and Comau the intelligence to have an open approach, so you can work on the training offer by modifying or developing it, that undoubtedly enriches it.

What did your company/organization and you deal with within the educational robotics initiative?

There are a series of data that say that we need to work a lot on the basic skills of children, because the surveys show a not exciting map of the level of educational achievement of our students as the spread of STEM and basic skills is low and, despite the fact that for many years we have been talking about digital skills, the maturity of the school system and out of school is not so advanced. Therefore, we started creating a series of empirical experiences of collaboration with the school system, on the one hand, and private on the other, by private we mean families, to offer, in some cases for free, in others with formulas providing edu-tech paths. When you think about edu-tech, that is, about educational paths linked to digital competences and basic STEM skills that see technology as an enabling tool, as a paradigm, the areas are substantially defined. There is definitely a whole theme related to programming, coding, there is definitely a whole theme related to electronics, there is one related to what can be called Internet of Things and then there is one that is the theme of robotics. So the added value, for example of the path that we have intercepted with Comau, which is not the only one to make robotic paths but it does it in such an excellent way, is not so much the fact that I teach you how to do a single operation with a robot, but I teach you to interact with the robot itself or, and in my opinion this is a qualifying element of the collaboration that we have started with Comau, is the fact that for example the experiences with e.DO allow you to work open source on the software that then again refers to the concepts of collaboration.

Interview 6 – Roberto Dentale (Asse4)

Name and Role

I belong to an organization called Asse4, it is a networking contact point between multiple companies. I specifically deal with the school area, everything concerning education. Asse4 is a network consortium, it is a type of aggregation of companies, within our structure there are social enterprises and social cooperatives, so for the issues we deal with we have a third-sector sensitivity. In Asse4, we are involved in the development of projects in schools covering different areas: we started with entrepreneurship education, with a program called studentslab, which is a project for the enhancement of excellence, working on soft skills, in particularly entrepreneurial skills. This is the project from which we were born then, over the years, a little for the market and a little for the need of innovation, we have expanded our types of projects, starting to deal with civic education. In fact, through the "Cittadella della Legalità" program, students have the opportunity to explore a number of issues related to legality with the involvement of several very important partners. These are the main projects for which we have been known in the world of school. When I talk about schools, I am referring in particular to secondary schools, so the age range is between 16 and 19 years, although in fact, in recent years, our projects have also been aimed at a lower range, primary and middle schools.

Short description of the educational robotics initiative (name, participants' description, place, time, goals and program, results)

We have been partners with Adecco for years, which has supported us in the implementation of our activities. In virtue of this relationship, some time ago we intercepted Comau and specifically all that concerns the e.DO laboratory, all the tools related to educational robotics. We had some meetings to better understand the use of e.DO, we went to Fondazione Agnelli where we could directly observe its use for schools. This allowed us to realize our project that we now run on three axes: us, Comau and Adecco. The project is very simple: it actually has the function of training people who can find a job outlet in Industry 4.0, taking into account all the new dynamics of work. Comau has designed and then shared the possibility to use two educational formats: one is called "Maintainer 4.0" and the other "Operator 4.0". These two courses are delivered to young people, some have already finished secondary school and are entering the world of work, others are already university students, however they are people interested in delving into this type of issues. The project is managed from a technical and training point of view by Comau, we are the training provider, while Adecco is interested in supporting this type of training in order to insert competent workers in the world of work. Companies are hungry and thirsty for people who have this kind of professionalism and competence. We are carrying out several projects with Comau: we have created an Academy at the Parthenope University in Naples, where we basically provide training courses. We are developing an alliance with them; the goal is then to extend ourselves on the territory with other partners that will give us the opportunity to intercept companies interested in the young people we are training.

What did your company/organization and you deal with within the educational robotics initiative?

We are talking about the link that is created between the student and the company, we are bringing them closer to the world of work on a specific request of companies, which is to have workers competent in Industry 4.0. On the one hand, there are the skills that the participants acquire using industrial robots, so e.DO or other robots that we have and that have been provided to us by Comau; in addition to this, then there is a whole company that is Adecco, in charge of inserting them in the world of work, which is no small thing. There is a good chance that these young people, once they finish the course, will find themselves hired in a context like Industry 4.0, where there is a huge demand and a potential worker, if capable, also has bargaining power.

Interview 7 – Marco Gioannini (Fondazione Agnelli)

Name and Role

For a long time, my main function has been that of Head of Communications, and this in any case defines a certain profile, after which, my second function is to follow the projects. So, I take a step back: in the past, the fundamental pillar of Fondazione Agnelli was research, now I would define the Foundation as an entity that moves on two pillars, the second of which is the planning with schools, planning that I would say is almost always, if not always, in connection with research. Therefore, we try to make the two things correspond; we do not always succeed completely, in the sense that we try to intervene with projects in schools where research tells us it is more needed and we draw from time-to-time elements that are useful to integrate research, especially in terms of school policies.

Short description of the educational robotics initiative (name, participants' description, place, time, goals and program, results)

Combo was realized in a laboratory at the Agnelli Foundation, that is, in an equipped area of the Foundation called Combolab. There every day, during the school year, every morning there was a class that could be of different grades: primary, middle and high school. It lasted three years and three months, so basically leaving out the very beginning periods, the very first weeks of the school year when it is more difficult that you are all still in the right place and the very last ones, the ones of the exams, we delivered about 180-190 sessions per year, I think 180 was the average. 180 sessions per year in a school year that in Italy is made of 205-206 days, so almost every day. Then sometimes you would have a class that did not arrive: once a class did not arrive because there had been a flood, things like that, but we always did it; unfortunately, Covid stopped us. It was carried out in the morning, it was always carried out by educators and facilitators from Comau, one or two depending on the situation, supported by the staff of the Agnelli Foundation for the more organizational and logistical issues, and of course all the preparatory logistics, the selection of classes, the contacts with the classes, understanding how many students arrived that morning, whether 15, 20 or 28, from which school they came, if for example, and this was true for everyone but especially in primary and secondary schools, there were children with disabilities who were obviously all welcomed, but we also tried to be prepared for the different situations. Therefore, the function of the Agnelli Foundation was to keep in touch with the schools, to talk to the teachers, to prepare the ground for the facilitators and then there was a constant interaction between us and the young people of Comau, who practically had a first X-ray or a first photograph of each school that allowed them to prepare the type of lesson to give that day. That is what we did. Let's say, instead, that the teachers' involvement in the workshop has been, day by day, more problematic in the sense that sometimes, even quite frequently, especially in the case of primary schools, there has been a quite active participation of teachers in following at least the proposal, the scanning of the day proposed by the educators of Comau and, I must say more frequently, we have seen people who have taken advantage of it to stay there and watch, chat with each other and so on.

What did your company/organization and you deal with within the educational robotics initiative?

From our research on the Italian education system, among the various critical issues emerged, quite understandable for anyone who barely knows the school system, a delay in teaching that affects a bit all the spectrum of school grades, but with an undoubted emphasis on secondary, something better is done in primary. We also had experience in our own Combo laboratories, and however the idea of an experiential teaching, "hands on" that not only puts, as they say, the hands in the dough but works from a point of view of the relationship between students and teachers, there is a strong degree of interaction and cooperation, so much so that the project I will talk about was substantially Cooperative Learning, it is not only that, but if I had to reduce it to a single category of teaching I would say Cooperative Learning. So the delay of the Italian school in didactics that we have studied, we have told, we have declined with respect to some topics, for example we have specifically studied the delay of didactics in middle school, just to say one, has led us, however, in this will that the Foundation has, because the will is a precise

Figure 2. Questions 1,2,3 - Overall section

SECTION A - OVERALL

1. **What value do you think the proposed initiative has for the PARTICIPANTS (what was it for, what did they bring home)? Tell an episode to exemplify the value you perceive.**
2. **What value do you think the proposed initiative has for the COMPANY/ORGANIZATION (what was it for, what did they bring home)? Tell an episode to exemplify the value you perceive.**
3. **What value do you think the proposed initiative has for the TERRITORY (e.g., scientific community, municipality, citizenship, institutions, families, etc.) (what was it for, what did they bring home)? Tell an episode to exemplify the value you perceive.**

mandate also from our Board of Directors, to try to organize projects with schools in which we experimented with modules of new didactics that had a value above all, how to say, emblematic. In other words, we don't want to generalize our specific experience, we want to try to generalize, or at least to suggest to the school system that some things can be done in a certain way and not in another, and that's certainly where the Combo workshop comes from. Then there are other needs, certainly the Foundation, being an entity primarily at national level, whose interventions and especially the indications of policy aspire to have national relevance, but has always the need to guard its territorial dimension, so, I mean, it is not nice to forget that we are in Turin and therefore it is important for us to have a network of relationships with the schools of Turin, with the stakeholders of Turin, so it was also important to try to do something for the schools of Turin actually trying, starting from our reality, to do something that was somehow exemplary. Combo is one of the examples, Combo lab with robotics, and then we get to the fundamental theme: a robotics that is not for itself, it is not robotics for coding, but it is robotics for learning, for learning even disciplinary knowledge, skills that are part of the curriculum and everyday life, and this is one of the examples.

Analysis of the Interviews

In this section of the chapter, authors will present the results of the interviews relating to the in-depth analysis of the different levels of value generated by the *e.DO Experience* project for Participants, Company and Territory, according to the perception of Comau managers and partners.

This part contains the responses of the interviewees to the questions showed in Figure 2.

Value for Participants

The level of value "for participants" is hypothesized to be supported by four macro-categories, namely "Skill Creation", "Value for New Citizenship", "Empowerment" and "Inspiration and Orienteering for Working".

For each of these, authors explored the areas and indicators that emerged during the interviews. They are presented below.

Skill Creation

The theme of "Skill Creation" was **one of the most recurring in the interviews**, often described as the main element of value for participants from both the managers' and partners' perspectives. This category of value refers to the possibility, through the *e.DO Experience*, of developing knowledge and **skills of technical, scholastic, work, relational nature,** and of **familiarization with new technologies**. Below are illustrated the main elements that emerged during the interviews, divided according to the skills that seem to constitute the "Skill Creation" area.

Robotic Skills

As told in the first pages through the voices of the various interviewees, the initiatives that have seen the use of e.DO are of diversified nature, aimed at different targets with different objectives and training content.

Some of these initiatives are in fact designed within special courses for robotic subjects, in which the robot becomes a real learning object to be known in its own right. In this case, **the connection with the development of the so called "Robotic Skills" appears immediate**, basically consisting in the ability to program and control a robot, to understand its potential and areas of use. Educational Robotics programs tend to address a relatively adult target, starting from students attending the last years of high school to young people who are entering the world of work or have already started their professional activity. These initiatives aim at the development of skills of robotic nature with the primary objective of training competent professionals in this field (Interview 3: "*Comau certifies that the student knows how to use and program an industrial robot and it is a certification that you put in the curriculum and you spend in the world of work* "), responding to the ever-growing market demand for this type of know-how.

Until now have been just considered initiatives designed and implemented precisely for the development of robotic skills, so it is to be hoped that they will generate a consistent result. But what happens in all the other cases, in the many initiatives in which e.DO is conceived as a tool to promote the learning of curricular subjects and soft skills? It is indeed true that most of the initiatives carried out by Comau for primary, middle and high school students, **consciously attribute to e.DO the role of a tool for learning**, as explained in this report:

"We deliver classes with e.DO but not robotics ones, this is a very important issue for e.DO, lessons with robotics. Robotics for us is not the purpose of what we do with e.DO, it is the tool. We start from e.DO as a tool, as an interesting tool, and then we convey other contents that are, if we talk about the school world, the curricular subjects. Then, however, this type of use allows an approach to robotics and new technologies, so it is a consequent advantage." (Interview 3, Manager).

However, the fact that e.DO assumes a role as a tool does not preclude, as emphasized by many interviewees, the possibility of conveying through it a **familiarization with robots and the development of skills to use it and relate to it.** It is not the primary goal of e.DO initiatives, however, it is a tool that was also designed and implemented to facilitate people to become familiar with robotics, in parallel to the teaching of curricular subjects, as argued in these verbalizations:

Interview 3: "It's a robot that Comau has developed to bring anyone, and I mean anyone, closer to the world of robotics, of new technologies, but also of mathematics and therefore of the subjects that are studied at school."

Interview 7: "When the child is younger is learning without even knowing it, things he or she does not even know; the child thinks to do the commutative property and instead finds out the understanding of something about robotics as well, so digital skills, let's call them in this way, and then growing up you put it all on the plate."

The hardware and the learning processes that are triggered make **e.DO a facilitator for learning, able to facilitate an initial familiarization with robotics and the development of skills in this area**.

In particular, it is reported as a critical element the fact that e.DO is controlled by a simple interface, which "*does not require knowledge of programming languages, but simple visual skills*"; this aspect is considered very useful to **reduce the fear and distance** that participants perceive towards a robot, helping them to consider it as a tool within their reach, as highlighted by the following interview excerpts:

Interview 3: "e.DO is controlled with an app on the tablet, so this helps participants becoming familiar with robotics without perceiving it as something scary, distant, difficult, rather something really within everyone's reach."

Interview 4: "Introducing young people to technology right away, making them understand that robotics is not something difficult, but something you can do with a tablet, is one of the big added values for participants."

Interview 5 "Once you had to go online to approach a robot, put on your shoes, there was a net between you and the robot, that was a difficult thing to manage, dangerous. Well, now I am going to put the robot on the table, I am going to program it with a tablet and in 20 minutes I can teach you how to program it, not with a programming language but with a visual language."

The type of learning processes that are triggered, considered as "active", "engaging", "exciting", "experiential", "hands-on", also accompany the **consolidation of the knowledge produced** and increase the possibility of **rapidly increasing one's skills to the point of achieving real "robotic stunts"**:

Interview 7: "Putting things into practice right away makes the learning process very fast and so you see, in our labs, how you start at the beginning from knowing nothing about a robot, to moving the robot in three seconds, to ending up doing stunts, excuse me for the metaphor, robot stunts."

Interview 2P: "If I have to tell you the value for the participants, I think that especially at the beginning the surprise effect was the discovery of the world of robotics, so you could see the faces, both children and adults, extremely excited by this strange thing that they started to discover there and that fascinated them so much."

The comparison with the interviewees allowed us to confirm what is illustrated in the literature: although many initiatives related to e.DO are conceived as educational robotics courses, in which the role of the robot is that of a tool for teaching curricular subjects and developing soft skills, its usefulness for the development of robotic skills seems undeniable. The distinction between robots as a tool and as learning objects is therefore not so clear-cut, since even just the fact of using e.DO as a tool to learn school subjects **helps participants to become familiar with robots, to understand their operating principles, their use and their potential**.

School Subject Skills

The second indicator of the "Skill Creation" area taken into consideration concerns the possibility of employing e.DO for learning school subjects of different nature. This element is also considered as particularly significant for value creation, specifically with respect to the possibility of promoting new forms of innovative teaching and active, inclusive, hands-on learning experiences.

The most immediate connection between e.DO and school disciplines appears immediately linked to the learning of STEM subjects, particularly mathematics. Comau, as well as many partners, has then started the development of training packages aimed at teaching other disciplines, such as biology, physics, moving also to the humanities with paths related to history, music and literature. From this point of view, several development spaces remain open for the application of e.DO to different disciplines. From the perspective of the interviewees, the use of e.DO appears to **facilitate learning in school disciplines for several reasons.**

First, the kind of hands-on experience allows the student to **understand the concrete uses he or she can make of knowledge**, that might otherwise seem abstract or notional, as suggested in this excerpt:

Interview 5: "A middle school student, at the end of a lesson, was talking to his professor and as he was leaving the class, he said to him "Professor, finally I understood why we study the Cartesian plane in school." How did he understand it? During our lesson, when we did mathematics with e.DO, we made him program the robot, but to program a robot you have to know X Y, the Cartesian plane, and therefore he made the contact with reality, he understood that what he studies in school is something he needs because one day, maybe also working with robots, e.DO, a robot, he will be able to control it because he knows X Y."

An additional characteristic of the *e.DO Experience*, namely "the strong emotional impact", guarantees the participants **a greater memory of the experience and of its contents**, consolidating the storage processes that are typical of long-term memory. In this case **the advantage will also be for the teacher** who, at school, will be able to offer a link to the students:

Interview 7: "From the point of view of the teacher, who maybe accompanies the students, it remains something to hold on to in order to recall certain concepts, so the teacher can tell the student "We went to see it, we did it...

how did you behave", everything: not only for the content, for instance math, but also for the behaviors, how they lived the experience with e.DO and how they used the acquired knowledge."

A further facilitating aspect emerges from the interviews, linked to the increasingly widespread difficulty of students in maintaining a **high concentration level**. Traditional didactics seems to struggle to intercept the needs of students and adapt the training method accordingly. Initiatives of educational robotics such as the *e.DO Experience*, thanks to **active experiences, challenge and collaboration**, seem to offer an answer in this sense, proposing **a training method that allows the learning of school subjects in innovative ways**. There are different types of e.DO packages proposed: in some cases, schools take part in "one shot" experiences lasting one day, in which students from various classes attend lessons prepared by facilitators from Comau or partner companies. In this case, the experience is single and focused on a specific topic. Another possibility is that organizations, associations and educational institutions create their own internal laboratories equipped with robots, develop training packages independently and use e.DO regularly as part of their curriculum. In this case, it is easier to observe the long-term effects of e.DO on the educational effectiveness and learning of school subjects, as the experience can be proposed continuously. In both cases, however, it can be seen that this type of experience allows students to change their predisposition during the learning process, making them more "receptive" to the notional content of the lesson. In fact, managers and partners report situations of **strong engagement of the students** who, in their own teachers' opinion, show a **higher level of attention** and **a different performance**, if compared to traditional classroom activities.

Interview 7P: " A student arrives home fatigued from school because actually the bird that passes outside the window attracts the same attention as the teacher, right?! So, in the end one is dazed and then arrives home with a headache and then we need to work on the motivational dimension with these kids. To expect them to pay attention to the lesson is like expecting the moon, and teachers need support on this, a tool like e.DO is perfect in this sense."

Interview 6: "It is very interesting to see the change in hierarchies from the classroom. Often teachers notice how students who make little at school, in our labs become instead top of the class."

e.DO is therefore a means to **facilitate the learning of school subject skills**, for the type of experience that is able to trigger, active, engaging, collaborative and experiential, as well as for the type of memory that creates in the participants thanks to the strong emotional impact. The use of e.DO offers an additional advantage, namely the fact that it **changes the students' predisposition towards learning processes.**

Working in the Future Skills

A considerable value element underlined by managers and partners consists in the possibility to develop, through the *e.DO Experience*, **enabling skills for the world of work of the future**, as highlighted by the words of a Comau manager:

Interview 1: "We try to prepare for the world to come, the industry of tomorrow, but it is already here today, isn't it? In the sense that we always talk about industry, manufacturing 4.0, the future, but in fact, guys, it is already here now. Let's talk about, 5.0 because now it is very close."

But what skills are needed to work in the world of work of the future, and which of them are implementable through the *e.DO Experience*? The analysis of the literature conducted in the previous chapters seems to be supported by the data collected during the qualitative phase. Managers and partners argue in fact that the areas of competence required concern elements of human relations, man-machine relations, ability to adapt quickly to changing and versatile contexts, to be engaged and motivated, able to collaborate, open to interdisciplinarity and innovation. Authors are referring to the results of a previous research (Fregnan, et al 2020) to classify these skills in seven capabilities: "Deal with Human", "Deal with Technology", "Agility", "Engagement", "Collaboration", "Interdisciplinarity"; "Innovation".

Regarding the item "Deal with Human", the usefulness of the *e.DO Experience* is linked to the development of the so called "soft skills" or "life skills"; if on the one hand this aspect is a real goal of the educational robotics initiatives under consideration, on the other hand it is a direct consequence of the ways in which the training is designed and conducted. Among the many "soft skills" identified in the literature, **those most elicited by the interviewees, as evidenced by the verbalizations below, concern the areas of teamworking, communication, problem solving, cooperation, role definition and task organization**:

Interview 6: "We also work a lot on soft skills, i.e., soft skills, organizational skills, cross-functional skills, it is definitely a valuable thing compared to what they are often offered instead."

Interview 3: "In addition to bringing STEM, i.e., raising the level of the ability of young people to be familiar with STEM, the thing I like, given that I am a humanist by origin, is that I was able to spice it up with all the soft skills and all the "real skills", i.e., not only STEM but also all the aspects of relationships and interpersonal dynamics."

Interview 4: "The different soft skills that you go and list from time to time when you have these types of experiences: collaboration, cooperation, teamwork, 'hands-on' and all of those things are absolutely true and very important."

Interview 2P: "The group work was another fundamental thing that e.DO pushed a lot and that the schools had started to propose and on which we saw big differences: there were schools already accustomed to which you could see the group that had dynamics, structured logics because they were already used to it and those who instead "one the hammer and the other the screwdriver", in short were not used to work in groups. From this point of view, e.DO was very useful in the workshops that we delivered."

Interview 1P: "The fact of working with robotics leads to highlight a very important competence in the tree of Life Skills, which are even broader than those of Gardner, there is the theme of collaboration, of team working, that is, practically working together on e.DO necessarily involves the fact that a group of children, a group of students works together and develops this type of competence, soft competence, let's say, that among other things is not only the school one, the hard one on the fact of knowing how to program a robot, but of working in a group".

Interview 5P: "That is, in our tasks, in our challenges, we work on communication, on team work, obviously, on problem solving, so we based the exercises on these transversal dynamics, on the division of roles, we give a task and the students have to organize themselves within the group in order to reach the goal in time."

Although the robot is considered as a learning tool and not the actual object of the *e.DO Experience* initiatives, another "Working in the Future Skill" identifiable in the interviews is the "Deal with Technology" competence, considered on the one hand as the ability to use and, in some cases, program a robot, on the other hand as a familiarization with Robots, understood as a founding technological element of Industry 4.0 and Digital Transformation. From the point of view of managers and partners, e.DO helps participants to **conceive the robot as something accessible, full of potential and with wide possibilities of use**, not as something to be afraid of:

Interview 7P: "With primary school children, using a robot even just to study curricular subjects can be a way to bring them closer to robotics and to start developing digital skills. It is not the actual object, but it inevitably brings you closer to the concept of Robot."

Interview 4P: "An element that is not curricular but, in my opinion, will become part of the things to do at school is the relationship between man and machine, therefore all the Human and Technology, so do not be afraid, the robot does not take away human job, but you have to learn to live with technologies and make the most out of them, therefore this is a bit, also in psychological terms, that is to question the relationship between man and machine, the sense of the e.DO Experience."

Interview 5: "There is a fundamental aspect that applies to both young and adults who come into contact with e.DO, which is the relationship with the robot. It is not a robotics course, but if you work with the robot, however, you learn to use it and you perceive it as an ally, not a threat."

Interview 4: "We also work on interfaces, i.e., how do you communicate with the machine, how have the languages of technologies changed, so that we become more and more communicative, right? How has the language of human-machine communication evolved."

The interviews also emphasized the competence identified in the literature as "Agility", understood in an adaptive sense and cognitive flexibility, as the ability to **easily adapt to new contexts, objectives and working conditions depending on the circumstances and opportunities** offered by the environment. In particular, e.DO is considered as **a "gym" to "train"**

this competence, as the experience that revolves around it provides that the student is in a new scenario, unfamiliar and called to respond to new tasks, to work in ways different from every day and relate to professionals other than the usual.

Interview 3P: "Then the whole theme of flexibility, agility, doing something you are not used to and having to adapt quickly, comes up."

Interview 2: "For example, many students are not used to working in a group, here they find themselves doing a task with others and not alone, achieving a shared goal. It is also required to develop an ability to adapt, to think differently than usual."

Interview 7P: "e.DO teaches that you can find many different solutions, that you can choose between faster or more effective solutions depending on what you have to do at that moment. After all, we are the first ones having to be flexible, to adapt every day to the class in front of us, I think it is normal that we then translate this into the different activities."

Interview 5: "The world has changed so much and continues to do so, and consequently the way of working. If you don't keep up with the times you are left out, to avoid this you have to adapt continuously and quickly. e.DO in this can help because on the one hand it allows you to learn how to use a tool that will be more and more widespread, but on the other, which is perhaps even more important, because it makes you understand that there are many ways to work and to find solutions."

Two additional competencies related to "Working in the Future Skills" emerge, such as "Engagement" and "Collaboration". Again, managers and partners see the type of experience as key to the development of these skills. From their perspective, in fact, although the content of the lessons is scholastic in nature, the *e.DO Experience* is supported by **collaborative logic and teamwork with a strongly emotional and experiential nature**. This operation allows the participant to **experience engaging moments of challenge and collaboration**, strengthening their skills in this sense:

Interview 4: "Collaboration is another great value, because the moment we can get these people and these students to collaborate, we really see that they grow and, in a very short amount of time, they learn quickly how to work together."

Interview 6P: "Maybe a student is taking a math class and learning the content, great. But what is around that? There is the ability to work together, with motivation and desire. You learn math and, more importantly, to work with others."

Interview 3P: "The companies we work with often tell us that in addition to technical expertise, which is increasingly rare, there is another thing that is difficult to find: the ability to work with others, to collaborate. We tried to respond to both these needs with an innovative project and there came e.DO."

Interview 5: "The feedback that teachers often give us is that students are participating much more, they are more attentive and involved than in traditional classroom lessons. As they say now, they are very 'engaged' in our labs and e.DO is instrumental in that."

Interview 2P: "The good success of e.DO was related to the fact that you finally collaborated, not that there was the genius of robotics inside, because it was not enough and the facilitator pushed from the beginning, because then, even in the game, the recognition was for the group, so the fastest group but that had not made everyone work was somehow penalized or not rewarded, as it would be in a situation of normality."

The *e.DO Experience* is designed with interdisciplinary logic, regardless of the school content, in fact, elements related to the robotic or mathematical field are always present. In this sense it is a vehicle for the **development of the "interdisciplinarity" competence**, since it inserts the student in this type of logic.

Interview 3: "Interdisciplinarity: another key word, besides versatility, it is really the interdisciplinarity of both e.DO as an object, because e.DO is a robot and it represents robotics, but it works because of math, you have to know formulas."

Interview 6: "It is interdisciplinary because it touches so many topics. Robot History touches technologies, touches history, touches you, that is, how to say, it also touches aspects of technological evolution that go beyond, touches programming that is more computer coding, because clearly the industrial revolutions bring you to add know-how. What I love most is Robot History, that is the history of industrial revolutions, so we do history with e.DO, I am doing history but our training is interdisciplinary, you cannot say it is purely curricular."

The "Working in the Future" competence that emerged during the interviews also concerns the value of "Innovation", in relation to aspects of technological development and teaching. In this sense, **innovation is understood in terms of values rather than competence**. These aspects will be explored more in detail in the "Value for Territory" section.

Interview 7: "Thanks to this didactic tool you are not only able to give value to the participants, to the companies, to the community but, in another sense, you are also able to give sense to innovation and research because it is a simple tool, it is a flexible tool, you can immediately go to the point of the situation, to make your own immediate concept, just because there is not a huge knowledge gap to enter, there is not an entrance badge that you have to pay to get to the point, you get to the point immediately and then from there you then grow up to develop your thesis."

5P interview: "It is an innovation to the way we do teaching and learning."

Interview 3:"It teaches that innovation is important, that they need to be able to innovate continuously."

The last "Working in the Future" skills' plan that emerged in the interviews responds to the heading "professional skills," understood as the development of technical skills, related in particular to robotics, enabling professional activity in Industry 4.0. One of the aims of the *e.DO Experience*, according to what managers and partners say, is to train competent professionals, able to enter the world of work and respond to the growing demand, by companies, for workers technically prepared in relation to new technologies. The great advantage identified lies in the fact that **e.DO takes up the concept of industrial robots**: by learning mathematics, students also learn how robots work:

Interview 2: "This is the value for the participant: you have the technology in front of you, which is the technology that empowers you to a 4.0 profession or a 4.0 career startup and, however, what I give you more of is the method which is 'know, learn and apply."

Interview 7P: "The real value that I think is left to the participants is that they learn to work with technologies, in this case robots, in the sense that they know how to program them and how to use them. In that sense, you are already forming professionals, people who are already oriented to the world of work and have a strong bargaining power in the 4.0 labor market."

Interview 1: "The big advantage is that e.DO can be seen as an industrial robot, so the students can learn math and at the same time how to work with the robot."

Interview 5P: "You train people who are capable of working, technically skilled, and who have developed their strengths. That is something that companies requests so much."

The "Skill Creation" area of value for Participants results to be **one of the most solicited by the participants**. The *e.DO Experience* is seen as **an opportunity for the creation of social value**, as it allows students to **develop technical and personal skills, related to the area of robotics, the development of school knowledge, as well as professional and enabling the 4.0 world of work**.

Value for New Citizenship

A second area related to the Value tor Participants that emerged during the interviews refers to the theoretical construct of "Value for New Citizenship", identified by the desk-based research in Chapter 3. Within this object, we find in the literature a series of indicators presented here: Challenging, Stepping out of Comfort Zone, Respecting Others, Exemplary and Consistent, Keeping Promises, Measuring Results Objectively. In this sense, the interviewees highlight, albeit in a mainly implicit form, how the *e.DO Experience* can be conceived as a way to develop in participants **a sense of responsible and generative citizenship**. Here are presented the results that emerged from the interviews in relation to the various indicators.

With regard to the "Challenge" item, the *e.DO Experience* represents a context in which to experience challenging situations and competition with oneself and with others. **Challenge is an essential element of these initiatives**, designed precisely with this in mind. Students therefore learn to work in contexts in which they are constantly challenged and in competition with others.

Interview 4: "Our idea of experience, of the e.DO Experience, is essentially to put both in collaboration but also in challenge people, a positive challenge, leading to a goal, generally when we decide the goal, we make it so that it can be achieved by many groups."

Interview 2P: "Through competition, virtuous processes are triggered. We talk about a different method to learn, but also a way to achieve the result. Competition must be positive, respectful of one's peers, the context and the other competitors. I think this is a nice value for the participants but also for the community, because they can bring these things in the classroom, in the family, if they play sports also in their team."

Interview 7: "We divide them into groups and create tasks that they have to compete with other teams, in the end everyone achieves the result. I have seen a lot of fair play situations, where different teams were helping each other or where they were congratulating each other at the end even though they did not win. I do not say they had lost, because in the end there is never a real loss and they take something home as well."

The item "Being Exemplary and Consistent" emerged independently only from a few respondents and in a mostly implicit form. However, it can be interpreted in different ways. On the one hand, the interviewees expressed how in the classroom **some students set themselves up as a model for their peers**; on the other hand, they expressed a deeper level pertaining to the sphere of meta-communication: in this sense, **are the entities that provide the experiences with e.DO that offer themselves as a model for the participants**, to show them an example of consistency "*Between what you say and what you do*". (Interview 7P)

Interview 4P: "So that person behaved in a certain way and you see that as a positive example, meaning that you take what we call best practices around the world."

Interview 7P: "We try to teach students to be flexible, but we are the first ones who have to be flexible. We have to be flexible based on the age of the students, their background."

Interview 5: "e.DO has developed with an agile approach, of continuous comparison with the user, but also with the different customers. It is clear that this becomes then a reference model that is also operationally translated into experience with e.DO."

Interview 4: "The participants are faced with an example of innovation, from which they can take inspiration for their life."

Interview 3: "It often happens that you see the student who is a bit more awkward, who according to the teacher in the classroom is one of the most oppositional ones, then in the workshop he helps his classmates in difficulty and becomes a positive example for others. Maybe it is the same student who makes fun of his classmates in class."

Participants are also encouraged by the *e.DO Experience* to step out of their comfort zone, experimenting with new and challenging activities, working in teams and achieving new goals.

This indicator appears to be related to **experiencing new modes of learning**, that are experiential and collaborative, as well as **experiencing new educational dynamics** and **interfacing with innovative technologies**, which they must learn to interface with before they can use them.

Interview 5: "The group work was another fundamental thing that e.DO pushed a lot and that the schools had started to propose, on which we saw great differences: there were schools already accustomed, so you could see the group that had dynamics, structured logics because they were already used to, then those who instead "one the hammer and the other the screwdriver", in short not accustomed to work in groups. From this point of view, e.DO was very useful in the workshops that we delivered".

Interview 2P: "You get to experience yourself in something new that you are definitely not used to. You have to recognize that it is something unfamiliar and you have to know how to overcome your fears a little bit, because you have to expose yourself in a different way."

Interview: "For a lot of students it was the first time they had a robot in front of them, we were asking them to do things they did not initially think they could do."

The indicator "Respect Others" is represented by respondents in terms of **equality** and **protection of the surrounding environment**. Specifically, it is emphasized that e.DO is particularly effective in **flattening differences among participants**, also in the case of disabilities or learning disabilities, reducing gender disparity in STEM fields, changing the learner-teacher paradigm and taking care of the resources available to them, while avoiding creating risks for others.

Interview 7: "You get 9 year old students who behave much better than a lot of adults; they manage to include their peers, when maybe in class they struggle because they are together every day, they disrupt because maybe they just have fits, problems, and that is when you realize how mature they already are, how grown they already are, and you help them because you do not have the constraints of the structure that you have in school, and that is when you can play it a little more."

Interview 5: "You put together kids who maybe live in very difficult contexts, you put together kids with many kinds of handicaps, obviously physical handicaps do not exist in these contexts, right?! So, we are all the same from that point of view."

Interview 5P: "The kid is holding an item of value, being empowered and treating it with care. There are also potentially dangerous items, so you have to put care into it for others as well."

In association with the item "Measuring Results Objectively", the interviewees consider several aspects: on the one hand there are real tasks to be achieved as a group, on the other hand the *e.DO Experience* is understood as a means of **measuring the quality of the group work that led to the achievement of the task**. It is then added the dimension of **awareness of the consequences** of one's actions, concretely measuring oneself with the results achieved:

Interview 7P: "There is another one that I think is very important in the digital era and that strongly differentiates an experience like Combo with robots from other totally digital experiences, that is the irreversibility of some causal actions, of some causal phenomena to which kids, and in this case there is no longer this big difference between younger and elder kids, seem not to be used anymore. There are moments beyond which an action is irreversible, if you have made a mistake you have to carry it with you, you cannot reset or have a second life as in role-playing games, that is, "you are dead, you are dead, you cannot reborn".

Interview 3: "At the end you come to a result, it can be good or not. You understand why it went well or why it went poorly, where you can improve and where you are already good."

Among the various indicators hypothesized, we did not identify clear references to the "Keeping Promises" item. The other theoretical assumptions seem to emerge from the interviews, albeit in an often-implicit form. The *e.DO Experience* seems to configure itself as a means for creating value in terms of value for new citizenship, with particular reference to the indicators "Challenging Oneself", "Going out of one's Comfort Zone" and "Respecting Others", depending on the logic with which the initiative is designed, the ways in which it is implemented and thanks to the characteristics of e.DO. **The items "Be Exemplary and Consistent" and "Keep Promises" seem to need further study**.

Empowerment

An additional area of the Value for Participants level that we hypothesized following the literature review is that of "Empowerment," which can be identified through the indicators of "Being Aware of one's Potential," "Developing Skills," and "Proposing New Solutions or Initiatives."

The interviews highlighted how **all of these aspects can be traced to the *e.DO Experience*, particularly the aspects related to awareness and skill development**. These two areas only partially address the field of robotics, which is certainly central, but also extend to issues of relational, social, learning strategies and curricular school subjects.

The interviewees underline how the *e.DO Experience* represents an innovative way to experience one's own potential in a different context from the ordinary one, capable of **eliciting individual characteristics** that otherwise, in more traditional situations, might not emerge and, consequently, might not be observed.

Interview 6: "It is not about creating something new, but about starting with the strengths that a person has, perhaps without even knowing them, and basing skill development on that."

Interview 7P: "Maybe a lot of students who think they are not capable in school or say they do not have the desire to do it, in labs instead start experimenting, and then they find out that 'Okay, this thing is something I can do'."

Interview 3: "A student becomes aware of many things: of his/her robotic skills, of his/her ability to work in a team, but also of the fact that maybe there are better ways to learn than others. Maybe not in an absolute sense, but there is no doubt that each of us has his/her own method and maybe the method that works for me does not work for you. But you have to be aware of that."

The indicator "Propose New Solutions or Alternatives" is made explicit in an interview, the usefulness of e.DO in this case concerns the fact that the **participant can choose between different alternatives and solutions**, learning to evaluate their effectiveness within the specific situation:

Interview 4: "So there is not just one solution, but there are a lot of them, and there is not necessarily the best one, but the best one in so many ways. So, breaking down these solutions and being able to distinguish them is definitely a great value that participants take home."

The "Empowerment" area is relevant to the creation of value through the *e.DO Experience*, in particular because, as suggested by the literature and confirmed by the interviewees, it creates for students **a context and a situation aimed at expressing their potential** and **becoming aware** of their limits and potential.

Inspiration and Orienteering for Working

The final area related to the Value for Participants level that authors hypothesized through the desk-based research is "Inspiration and Orienteering for Working," expressed by the indicators "Expanding Knowledge," "Discovering Passions" and "Identifying Study and Career Paths."

The interviews show that both managers and partners tend to view this dimension predominantly in terms of **discovering a passion for robotics and a desire to explore this topic** further at school or in their own work environment.

Interview 3: "So, what is the added value for participants? The realization that robotics is something simple, absolutely accessible. Somehow the student who comes into our workshops says, "But, you know what? Maybe it is something that I can be interested in, that I can be passionate about", and so we direct him or her towards a path of study of that type."

Interview 3: "I have so many examples of students that after our workshops, in a very interested way, came to me when I had finished and I was setting up my workshop, saying "That was very interesting, but what do I have to study to do what you do?", and your eyes shine."

Interview 4: "Our focus is primarily on students we call vocational, which are the schools interested in robotics, interested in using robots; many times, these schools have done a lot of theory but they have never seen a robot in practice, or have never tried something concrete."

Interview 2P: "Then, if I have to tell you the value for the participants, I think especially at the beginning the surprise effect was the discovery of the world, let's say of robotics, so you saw the faces, both of the young ones and adults, extremely excited by this strange thing that they were starting to discover there and that fascinated them so much."

Value for Company

In this section are presented the different areas that structure the Value for the Company level generated by the *e.DO Experience* propject. The theoretical hypotheses authors want to test consist of the assumption that, through Educational Robotics initiatives, companies like Comau or its partners have generated value in terms of Core Business/Mission, Brand, Reputation, Employee Branding, CSR, External Innovation Potential and Employee Exposure.

Core Business/Mission

A first advantage immediately observable through the interviews is the generation in terms of Core Business/Mission. Comau, as well as several of its partners, is in fact a company that, as such, must rightly respond to business needs. In this sense, the first advantage that the interviewees identify is related to aspects of **positioning in the education** market. The various companies following the *e.DO Experience* have established themselves as **solid references** within this sector. In the case of Comau, this new branch of business has culminated in the creation of a new business segment precisely dedicated to education. Subsequently, a further element of value for the company in terms of Core Business is linked to product development. In particular, great importance is attributed to what Comau managers define as the "agile approach" through which, following continuous comparisons with customers, teachers and end-users, it has been possible to **develop a product capable of responding to the needs of the market and of intercepting its demands**.

Interview 3: "The initial idea of e.DO was actually a robot in kit format, that would come in kit form with a briefcase and that could be assembled and then used by the same people. That was an initial idea. In parallel to this idea of e.DO in kit to be assembled, there was the other wing that was the e.DO already assembled for the world of education, ready to be brought in schools. Then the business has led more towards the second direction that I described, so the idea of e.DO in kit to be assembled and used was a bit abandoned. But this also helps to understand Comau's approach, which was, if you like, a very agile approach, almost like a start-up, because this e.DO has, and I say it because I have experienced it all, an agile approach to the product. Why do I say this? Because it was developed, then it did not continue the development until the end, but it was immediately brought on the test benches, directly in schools, and now I will tell you how it was tested, a lot of feedback was collected, it was then brought back to the design stages and this feedback was implemented. Then it followed, I make this circular gesture because it is actually an agile approach, so it did not go from an idea to the final product, but there were many steps of testing with the customer, with the recipient, feedback, subsequent improvements that led to a better product."

Interview 5: "There is definitely an element of value in terms of business, without which a company cannot stand. It makes sense that this is a product and service package that is highly attractive in the marketplace, because in this area e.DO represents an excellence."

Interview 6: "You have opened a new market, new languages, new ways of communicating, in my opinion certainly at the level of capillarity we have reached, this new experience has reached markets that we would have never imagined it could reach."

Interview 3: "Now we exist as a business segment, that is, we are Comau Consultancy & Education. That is, the value to the company we have created in four years, an identity for ourselves, what we have created with results."

Interview 4P: "The first advantage for us as a company is definitely in terms of market positioning, we are a reference point within our sector".

Interview 3P: "From the point of view of the positioning of our brand, it was useful, but also from the point of view of business opportunities that have been created, because then, through European projects, possibilities can arise for the development of the Digital Innovation Hub that can also be translated into new hires".

The *e.DO Experience* offers companies, both partners and Comau, the opportunity to develop their Core Business/Mission by **opening up to new markets and strategic positioning choices**, as well as through **product development strategies that increase their market appeal**.

Brand

The Value for Company that is generated is understood by respondents in terms of **Brand development,** with respect to its knowledge and awareness. As far as Comau is concerned, the *e.DO Experience* allows to **modify the perception of the Brand from the outside**, leading to conceive the company as a world reference not only in the field of robotics, but also of Education.

The advantage in terms of branding also concerns the **diffusion of the Brand**, making the company better known on the national and international scene.

Interview 7: "The company certainly takes home, on the one hand, the fact that it becomes a worldwide reference for education, and especially of education not only regarding robotics, but it makes robotics become a means to learn many things".

Interview 6: "On the other hand, as far as the external part is concerned, certainly all the brand awareness that we carry on, in the sense that we are a small part compared to Comau's business but with a great media impact, for example, so it ends up on the news because you do something innovative, new, and you have customers who are already accustomed to communication."

Interview 3: "Again, it is a tool to do this new mode of training for the school world, and we went to Brazil with one of these centers. I personally went, a beautiful initiative, because FCA in Brazil bought one of these laboratories and gave it to a poor school, to a school not too lucky, so I went to train Brazilian teachers who were very grateful for this opportunity, because we gave their school a technological laboratory and it increased the social value of that center."

Interview 7: "I still find customers who often tell me 'Eh, but I am from Arezzo, I have never heard of Comau. Can you describe it for me?'; when I tell them that we have more than 9,000 employees in 14 countries around the world, they say 'Gee, how could I not know that?' So, it is a branding effort that is certainly fundamental, regardless of whether there is an economic return, it is already a value."

Reputation

A further element of organizational value concerns the area of Corporate Reputation, since Educational Robotics initiatives generate **new images, perceptions and representations** that are attributed by the outside world, not in terms of Brand, but **in terms of values and reputation**. In particular, the positive elements in terms of reputation that are identified by managers and partners, concern the possibility, both for Comau and for the partners, of being conceived from the outside as **entities really interested in the creation of a technological culture marked by development and innovation**: it is not just a matter of "selling robots" with business objectives, but of an **active commitment to contribute to the development of the community and digital citizenship**.

Interview 5: "I think the elements of Branding, Employer Branding, Corporate Reputation, Corporate Social Responsibility are evident, so it is all about image and perceptions that are attached to the company from the outside."

Interview 3P: "On the other hand, I think that the real difference, and on this we have to give credit to Comau, that actually the effort they put in creating the training packages, that is the philosophy behind it, is in my opinion the real innovation: you can understand that at the end they are not interested in placing robots and e.DO, that it is not this; you can see that there is something that other automation companies at the moment are not proposing, a real awareness that robotics must be taught from a younger age, and I think the team of guys who worked on the training packages also shows a passion that I could find in their offer; it is not the usual re-proposition of things already seen, I must say that there is an air of renewal, it is perceived precisely the novelty. Then, another aspect that in my opinion is not at all secondary, is the fact that Comau has chosen to use open source technologies and therefore the fact that inside e.DO a Raspberry has been installed; the fact that if one wanted to, one could disassemble that robot, rebuild it, extend it, add some functions, and they expose all the manuals themselves, the technical documentation; for a company, just because it sells robots, it was not taken for granted that they would do this kind of reasoning".

Interview 2P: "The territory sees us as a positive reality that has given work to so many people, and let's say, that the typical issues such as pollution, accidents, etc. are minor issues if compared to those experienced in other territories; so we have a positive value and positive reputation already of our own, the fact that then there is a Foundation interprets, and makes evident and transparent, all these values."

Employee Branding

A further dimension of Value for Company lies in the possibility, through Educational Robotics initiatives, to **attract potential workers for the future**. The strong emotional impact of the experience, the level of technological development and the strong engagement implied by the *e.DO Experience*, create **a strong link between the participant's memory and the company**. In this scenario, according to interviewees, it is possible to **attract potential workers** (likely already skilled in STEM and robotics subjects) for the future.

Interview 4: "A student who experiences so many hours of training with e.DO and therefore does so many hours of training with the Comau logo, maybe as he grows up, he will remember that he did training with Comau, so it is a nice return for Comau. Again, as I was telling you, in these four years in which we have been operating, we have met thousands, tens of thousands of students, and tens of thousands of students who grow up and see the Comau logo may remember the time they did the workshops with e.DO, so it is a good return for the company."

Formal CSR

The item "Formal Corporate Social Responsibility" **did not emerge during the interviews**, the perception we had is that both managers and partners do not see the *e.DO Experience* as a means of responding to regulatory constraints and obligations, but rather as **a real commitment** to community and territorial development.

CSR True Committment

The dimension of "Corporate Social Responsibility" seems to be expressed by the *e.DO Experience* in terms of **real commitment to the community**, not as a regulatory obligation: it is a real "giving back to the population", to share expertise, learning methods, innovative tools and know-how to guide a growth generated by a virtuous exchange:

Interview 2: "Sometimes I also work in areas where Comau does not exist, but the projects that we bring, let's say, if they are bought by foundations and often by bank foundations, by companies, it is clear that they do it for a theme of Corporate Social Responsibility and then make them available to their territories, hardly the foundations that buy these projects then resell them, often they do it to make available what they have acquired, so technology and training, they make it available to the people of the territory. This has happened in Northern Europe, in Russia, in other areas where these projects are brought and it is clear that they are positive for the territory because, as you can imagine, they create a value, they create knowledge, they create skills for people, for citizens, for local people."

Interview 5: "These are simple things, however if we do this digital culture as Social Responsibility, as giving back to the population, this thing increases the quality of life, aligns and also makes you happier if you think about it, because it is not something I do not understand, it is something that 'Whatever, I do not have to be an expert to do it'."

Open Innovation

Partners and Managers consider the *e.DO Experience* as **an opportunity to gather further ideas for development and innovation from the outside**:

if on the one hand these concern the aspects of product development thanks to an agile approach, on the other hand, the collaboration and exchange with teachers, customers and users has also allowed to develop more flexible and effective training methods.

Interview 7P: "If you do not engage with the outside world, you risk missing out on all the cues for improvement that you could not see on your own. Collaborating with teachers during planning, for example, allowed us to develop individual initiatives based on the class we were facing, and this was key to build the experience."

Interview 2: "We listened to the market, we grasped the need to combine the needs of educational innovation with the development of job skills, and this allowed us to develop an innovative product: we were among the first to employ a robot with an industrial format in an educational context."

Employee's Exposure

The last area of the Value for Company level is aimed at those who are part of the organizations that deliver the *e.DO Experience*, for instance both Comau and the partner companies. Recognition occurs on several levels, the first for the facilitators who, through contact with the student, feel a strong utility in their work; a second level of recognition comes directly from the company. In the case of Comau, for example, the creation of a new company segment was experienced by the staff as **a form of recognition and as a strengthening of their sense of belonging**.

Interview 3: "Also because the birth of the Consultancy & Education segment of Comau is recent news, so there is a good motivational lever, I am speaking on a personal level. I mean, I am very happy about this because I started when nothing existed, there were two of us in an office, and four years later I look at what it has become and we are now about twenty people who rotate on the education business, we have even had an official recognition because we exist at corporate level. That is, colleagues who look at us can finally say, "Hey, here we are the Consultancy & Education people." This is an internal value for Comau, I see it as a value for the company."

Interview 5: "I think that what is perhaps even richer in terms of value for the company is the issue related to 'upskilling', to contribute to the development of the company through these initiatives by making managers even more competent and able to dialogue with those who arrive now and meanwhile have a common understanding, at least some elements of common language, the desire to learn; sometimes those who arrive, maybe who do not know anything about your work, but look like people who know a lot, maybe you know a lot too, but meanwhile they stall to hear. So, it is a change in mentality."

Interview 6: "In my opinion there was a big internal impact, of real change, that is, of kick-starting a process of internal change when change is still going on."

Value for Territory

In this section are presented the different areas that structure the Value for Territory level, generated by the *e.DO Experience* project. The theoretical hypotheses that authors want to verify consist in the assumption that, through Educational Robotics initiatives, it is possible **to create a value for the territory according to generative logic, aimed at the creation and dissemination of a digital culture, educational innovation, the creation of ecosystems within the territory and communities and, finally, the sharing of knowledge and territory with the community**.

Creating Culture Dissemination

Speaking of creating Value for Territory, the logic of generativity comes into play, of sharing with the territory and the community with which it is inserted elements of the culture of robotics and innovation, enabling to live and work in a 4.0 world. The process of cultural dissemination described by interviewees occurs in multiple steps and through multiple agents.

- First of all, the **participants**, the direct target of the *e.DO Experience* initiatives, are exposed to the corporate expertise of Comau, a leading company in the field of robotics. The student represents a first vehicle for transporting cultural elements, as he/she can bring these elements into his/her own family context and, in a broader sense, into everyday life.
- A next step involves **teachers**, who do not belong to the generation of digital natives and, for the majority, are inexperienced in this type of issue. Their involvement is seen as fundamental in order to create

continuity with respect to the *e.DO Experience* initiatives even once the student has left the lab. Managers and partners also stress the importance of transmitting a culture of innovation, robotics and teaching to these actors, who are often less experienced than their students.

- The last step consists in the creation of a **network** with the territory through the development of relationships with diversified partners, which may be institutional bodies, schools, associations, networks of companies, etc.; the objective is to provide the territory with the means to carry out educational robotics initiatives independently. This result can be achieved, on the one hand, through the creation of robotics laboratories in schools or on the premises of the various partners and, on the other hand, through the "train the trainer" model whereby Comau, in addition to providing the technological devices, makes its technical and design know-how available to the territory. The partners are trained to be autonomous in the design and delivery of *e.DO Experience* initiatives customized and targeted to specific content.

Interview 2: "Thinking of going to schools but also to organizations, to companies to bring this project but also bring our expertise on topics that are precisely STEM, on the one hand, but then, as I said, the project has also matured, has expanded and has also taken into consideration topics closer to the industrial language; well, this is perceived as a value not only for those who participate in the programs, but also for the whole territory and the communities they belong to."

Interview 3: "They are creating, and I am going right in the territorial direction, they are creating poles, at national level above all, of learning centers with robotics sponsored by Comau but also by other entities, because the customers are others, that therefore increase the range of action for [...] that is, the territory is enriched with training centers with robotics that are able to proceed autonomously."

Interview 5: "The concept of inevitability comes to mind, for instance. a company like Comau thinks it has something to give because it has set out to reflect, there is a group of people who have set out to reflect, and said 'We have things to give, it is inevitable that I will give these things to you, if I do not give them to you, these things will happen anyway and you will be out', because we realized that the train goes by, it goes on and there is someone else who takes it and it is inevitable, either you get on or you do not, if you get on well or not well, both as a company and as a person, you will stay behind, you can see if another train goes by, go to the back carriage or stay on foot. So, there is a bit of a concept of inevitability in this thing here, so long as things happen nothing stops. We are not indispensable. Things happen anyway, Digital Transformation happens, so I would also value the ability to get on that train."

Interview 7P: "Educational robotics initiatives are also a means to bring teachers closer to the world of new technologies. The goal is to try to instill a few drops of somewhat new, of modern thinking, so that the teacher takes it home and continues to work on it."

Interview 6: "Around Comau, in my opinion, it has been created what have succeeded with our business model, which is not that of direct training but is that of "train the trainer", so it is like 'I entrust you with my know-how, I entrust you with my robots and I entrust you with a mission that is to bring STEM in the territory with this mode'. This is a big sounding board in the territory, so then, as if to say, 'I entrust you with my value and you have to make it sprout'. This in my opinion has been a functional business model for the growth of local communities as well."

Interview 1: "If these young people will become part of the industrial supply chain, they will manage and drive processes in such a way, and so we also talk a lot about sustainability, our discussions are also full of green, everything today is digital & green and green & digital."

Digital Transformation is happening and the radical changes it brings are inevitable: sharing with the territory a culture of innovation and robotics conveyed by the *e.DO Experience* **allows the community to develop those elements of digital citizenship able to respond to the social and work challenges of the 4.0 world.**

Educational Innovation

A further element of value for the area is always linked to a theme of cultural development, in this case linked to the culture of educational innovation. The *e.DO Experience* represents **a new educational model** from which to take inspiration and based on the principles of educational robotics. It is proposed a change in the paradigm of teaching, based on Upskilling and the possibility to learn while having fun, through active and engaging experiences of learning by doing.

Educational innovation impacts on several elements that we are going to describe in the words of the interviewees:

- New methodology for teaching school subjects and interdisciplinarity: (Interview 2) "*Teachers, educators, trainers identify in this methodological approach and this robotic arm a different way and methodology to approach topics, to interface with their students on topics such as precisely mathematics, robotics, programming that normally are told; apart from programming that is not common in schools, for example going to universities, in polytechnics, it is common in some training institutes, in some training centers and in some technical institutes, so this new concept of joining heavy robotics, let me say heavy, let me say it lightly, heavy robotics but, together with a series of topics, of subjects that are the curricular ones, it really creates a disruption, as you would say these days.*"
- Learning by doing: (interview 2P) "*The second thing was the application, for instance they realized that those absurd things they learned at school by heart were actually useful to do the thing they wanted to do, which was to move the robot and make it do what they wanted it to do; so, discovering that mathematics had a different function that was interesting too, it was another extremely relevant aspect when they sat down. It was the discovery of a huge world that opened up essentially endless possibilities, tied very much to an approach that is not that of mnemonic study but of proof, that is, 'I am going to stand there and try'. It does not work? I do not just say 'Wait, I will check the algorithm', I press this key, I see this other one, and then I say 'Oh, I pressed this key, it is true, because I put the abscissa and the ordinate' and that was extremely powerful for the students we saw there.*"
- New classroom modalities: (interview 3) "*What I think that what is useful is that e.DO overturns some classroom dynamics: the lesson with e.DO is not a standard lesson of the teacher-teacher-frontal lesson type. No, e.DO allows you to do different activities, more workshop-like. We have obviously removed the chairs from our laboratories, there is no chair, with e.DO we are not sitting, absolutely not. So, the teachers who have observed the lessons with e.DO have seen different classroom modes, but a very important thing that some teachers have*

taken home is that they have seen not only a different classroom mode, but they have seen a different response from their students."

- Role of the teacher: (interview 7P) *"I believe that the values brought by the teachers were basically two: a more general one, another perhaps more practical and operational. The first, more general, was to define some curricular areas that were particularly significant for them, in which they saw, once they understood what e.DO could do, and this was obviously explained to them by their colleagues at Comau, 'Well, this is something that could help us do that stuff better', this is the first thing. The second thing, and there they were really useful, because at Comau we did not realize it either, they helped us, once the basic module was defined, to have some information, some principles of flexibility, which then the educators who daily did the lessons adopted, to adapt to different ages and different classes, because sometimes you had a third grade and the other time you had a fifth, and the universe already changes between 8 and 10 years old; but, even talking about the secondary school, you could have some scientific fifth year students, maybe even from the computer science curriculum, that is, people who maybe already imagined going to do engineering or computer science and therefore had a lot of analytical skills, in short, they could teach most of the others, but you could also have on the one hand kids from the third year of the classical school, who frankly maybe had not even arrived at trigonometry, assuming that it is still done there."*
- Changing classroom dynamics: (interview 7) "*Some teachers have confessed to us that some classroom hierarchies have been overturned, in the sense that students who were perhaps less brilliant in the traditional classroom, with this new, different classroom mode, stood out. So, the teacher began to wonder 'Ah, but maybe it is just the approach I have in managing a lesson that impacts the response of some students', and then they took home the fact that maybe by changing the classroom mode, the response of the students could also be different. In our workshops with e.DO we do not give traditional frontal lessons, we put the participants in groups, each group has a table with five students with e.DO in the middle and they collaborate, they learn, they challenge each other; there is also a theme of challenge, constructive of course, healthy competition.*"

The theme of educational innovation was **one of the most recurring during the interviews**; managers and partners attributed to this dimension a very significant weight in terms of value creation for the territory. The value lies in responding to the needs of the territory to adapt the Italian training and educational model to the typical challenges of Digital Transformation.

Creating Ecosystems

A further element of territorial value lies in the **creation of networks and alliances** between the various players in a territory or a community. Through the development of these relationships, ecosystems are structured so that the various elements are interdependent and all indispensable to the functioning of the system. As stressed by both managers and partners, **the strength of the ecosystem lies in the active participation of Comau**, of the partner companies and, of course, of the students. Comau has set the structural conditions for the creation of value, i.e., on the one hand, the technologies and, on the other, its own know-how and expertise; this step is a necessary but not sufficient condition, because the partners have been indispensable for intercepting the local community and broadening the resonance of the initiative. Teachers were then fundamental to translate the initiative in function of the final recipients, without whom it would have been physically impossible to start the classroom activities. The value generated for the territory by the creation of ecosystems concerns the concepts of exchange, contamination of cultures, knowledge and experience. Then there is the value of the network, of being able to reach a wide territorial horizon and to create continuity over time.

Interview 2: "There is the value of knowledge and technology, as well as the knowledge of technology, but not only that, then there is the value of the knowledge of people, the value of the network that is created, the value of the ecosystem in which you are immersed and which allows you to broaden your horizons, the company's horizons. In a system, in an ecosystem, there is also a little bit of this exchange concept, of sharing, so it is important to underline that not only we are carriers of values, knowledge and image, because anyway Comau is a strong brand also in some areas abroad, but it is also important to say that this value is fed and increases thanks to the ecosystem."

Interview 4: "The aggregation of so many members, because participants, companies and communities have always, since the beginning of time, been interested in the educational world and see value within that, so I actually envision people being able to come into communication from all sides within this square. That also means a big exchange of cultures, of impressions."

Interview 4: "Positive contamination, and I learn a lot from people in other areas, as they probably learn from me, because my path is very technological, very directed towards the world of robotics, so there is a strong exchange."

The creation of ecosystems seems to be, in the interviews, **a central aspect of value generation** even if, among the different dimensions, it was probably one of the less explicit. One hypothesis is that this aspect may have escaped the awareness of the interviewees as structural and foundational: in fact, the creation of ecosystems seems to be considered as a means of creating value rather than an actual result of experience: "*All of this could not have been done one without the other*" (Interview 7P).

Creation of Skills Required by the World of Work

This area of the Value for Territory level is based on the **generative principle of capacitation**, understood as "enabling" people, creating the conditions to face the challenges and seize the opportunities introduced by Digital Transformation. The creation of skills, sharing of know-how and knowledge are conceived by interviewees as a value for the actual territory in terms of "capacitation". The *e.DO Experience* is a means of **providing enabling technical knowledge and skills for living and working in a 4.0 world, related to STEM, soft skills and robotics**.

Interview 2: "It is quite unusual, other competitors do it, there are some of them, but what characterizes our project, Comau's project? It is really the synthesis of these two souls, it is our characteristic of being used to a heavy industrial company that has normally always dealt with automotive and now is also looking at the world of General Industry, working on the design of training packages that communicate with the robotic arm, with the small e.DO, but that also communicate value, both in technological terms but also in terms of soft skills, to those in front of us: young people, professionals and the teachers themselves, because this should be emphasized."

Interview 2: "We try to prepare for the world to come, the industry of tomorrow, but it is already here today, isn't it? In the sense that we always refer to industry, manufacturing 4.0, the future, but in fact, guys, it is already here now. We also talk about 5.0, because it is very close, and so in that kind of the sense."

Figure 3. "Back to the Future" film

Interview 3: "Comau really attests that that student knows how to use and program an industrial robot and it is a certification that you put on the resume and you can spend in the world of work."

Interview 4: "But there was incredible success at the beginning, precisely because of this: because our strong value proposition was to bring the world of industry, the professional world, right from the start inside the school and therefore, even today, we configure ourselves precisely as a link, the connection between the school and the enterprise, and there in the middle we can really have our say."

Interview 5: "You are a capable person, so let's upskill a little bit, add a piece, and continue to be a capable person. So that is Social Responsibility, capacitating people a little bit."

Interview 7: "Obviously all the technical skills, the way of thinking, the mindset, that is another thing that you can easily share with the public and it is one of the most valuable things that you can find in a big company, because that is what makes the difference: how you think, how you approach problems, how you solve them; you set the example and maybe someone learns something, not necessarily copying or buying, but simply making one's own concepts after having seen you doing it during a training, any delivery, a trade show. So, a little bit in that sense, it tests your skills of all kinds, not just the technical ones that many people may have."

The Images: An Overview

In this section of the chapter, are presented some images that were narrated by the interviewees, explaining their meaning based on their accounts. We cannot speak here of a true semiotic analysis, as we try to limit our interpretation and instead to **provide the perspective of the interviewee.** *In fact, during*

Figure 4. Matryoshka

the interview, participants were asked to answer the following question: **"***What is the image (concept) that best represents the value / values you have identified?". Here are the most significant images.*

"An image can be represented by a title that we always use, which is the "working in the future" title, actually, if I think of an image, I have in mind the "Back to the Future" movie one.

Figure 5. Agora

If we talk about the value for the participants, we certainly transmit the passion and the richness of what is initially seen as a threat to man, which is the technology, and in this case we have tried to transform it into a captivating figure that becomes your companion, your accomplice, an accomplice that you will carry with you, especially for those who are lucky enough, but I think we all will be at the end, because it is one thing to think about, just the robot as the shape of the robotic arm, and then you are rightly positioned in a more industrial environment but, where you can understand that this robot represents technology at 360 degrees; just think about this situation, you and me in front of a screen, we already have technology in the middle of the human relationship, and in fact today, with all these great limitations, even our brains have been very upset in perceiving people in a completely different way, with a coldness to which we were not accustomed. So, I think that at a value level the big thing is this, bringing elements of the future into the present."

"What this evokes me in a narrative key is a matryoshka: a series of dolls inside dolls or boxes inside boxes, right?! This is the image that comes to my mind, because let's say that the educational robotics project is a project that is complementary to a series of topics and technologies, they are complementary to each other, but the little box can also live on its own. The little boxes are, as I was saying, the various project layers, they are the technology, then the hardware, the software, I mean the hardware, for instance the arm, the software that is inside that makes it move, they are all boxes one inside the other, the training, the educational packages. This is a project that contains several projects, they are souls that are one inside the other and they do

Figure 6. Fish that comes out of the bubble

Figure 7. Grandpa and grandson

not have to live necessarily on their own, but if you put them together, they make a bang and it becomes a large container of experiences, knowledge, skills, methodological approaches, training paths. It reminds me of sharing, of being able to put together so many things that form a complete whole and it will never finish, there will always be another box that contains another box, because it is a project that evolves."

"As an evocative image, let's say that in my conception, when I think of education, it is definitely the idea of a sort of agora, a square where we can make possible the aggregation of many members; participants, companies and communities have always, since the beginning of time, been interested in the educational world and seen value in it, so I actually imagine people who can communicate from all sides within this square. I come from the world of engineering and what really excited me about this adventure was to enter a world that is not mine, to go outside my mental conceptions as an engineer and actually see how other cultures, because they are essentially cultures, can blend into my background, and I see a lot of complementarities. Positive contamination I would say, and I learn a lot from people from other areas, as they probably learn from me, because my path is very technological, very directed towards the world of robotics, so there is a strong exchange and the agora is representative in that sense."

"The evocative image is a bit like the one that you see many times when you think about taking a leap into the void, which is that little fish that comes out of the bubble because, this is very interesting, when I think of education in Comau, so the big robotics company that for years has led the industry with its ideas, with its concepts of robotics, now also tries to get out of its pattern,

Figure 8. Kaleidoscope

out of its comfort zone, trying to challenge other areas. And it is precisely in that context that all the good things come out, on the one hand, as well as all the gaps that need to be filled in order to reach that type of solution, that type of customer, which is essentially very different from the usual ones. We have certainly filled those gaps over the years, it is now a few years since this project started, it began as a sort of start-up, but then it really became a pillar within the company and we immediately filled the gaps with this contamination. Comau was able to bring in new people, new personalities, and found the skills within the company that could understand this change, because it is not easy to talk to people who have been doing a certain job for years and get them to change their thinking completely. Fortunately, Comau has adapted and is still adapting, because we are halfway through the process, we certainly have not arrived yet, there is a long way to go, but let's say that the train has left."

"The first thing that comes to mind is a grandpa explaining something to his grandson using the tablet, so they are talking and grandpa is not just the one giving candies or telling boring things about the last World War. I do not mean the grandfather as the symbol of a person over 60 and who might be among those who have let themselves been pushed aside a bit, or who have just raised their heads after 40 and have a lot to say but have self-excluded themselves, living a life that is no longer in harmony with what is around. In this paradigm, the grandparent is a bearer of experience and knowledge, but at the same time who has been able to "capacitate" himself.

Figure 9. Face of a child

So, if we go social, one representative image might be this one. The grandfather himself who, in order to explain something to his grandson, takes his tablet and explains it to him. If we transpose it to the company context, it is the 50-year-old manager who still has so much to say, so much to do; it is the symbolic working of the little robot that becomes 'I am going to show you things using a digital language' and, without even realizing it, after a while you would say 'But... I am using a robot', and at this point we discuss together about what has happened."

"The image of a kaleidoscope comes to my mind: what we do is aimed basically at talking to the heart of the kids to bring out the inclination of the

Figure 10. Basketball Court

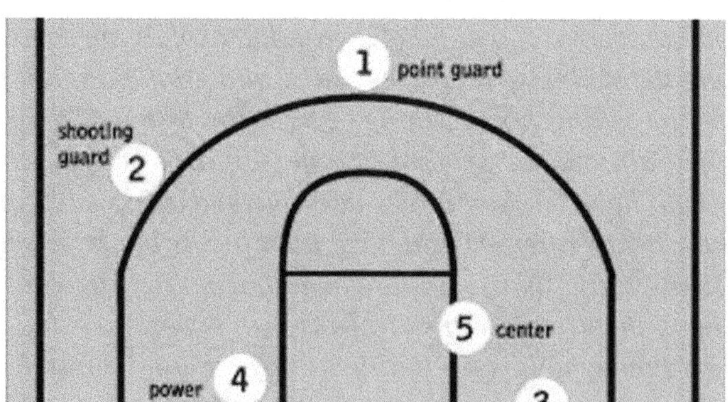

Figure 11. Villa Doria D'Angri

younger ones, the middle school kids, the primary school kids, but then it also continues for the elder students, high school guys, to make sure that these can go on over time, through robotics, coding and technologies in general, experiment and, through a more playful approach, train a little what is the 'kaleidoscope' that everyone has in personal skills, bringing them out and then making the right choice, more than right, a more enlightened one, more oriented than what is their own path of study, their career studies. We have been working for years through a method, a methodology that is inclusive and active for young people starting from the laboratory dimension, we do both school and vocational training, borrowing a lot of experience from vocational training in school, right? And basically, it is the method that changes, that is, the school transmits content, knowledge; vocational training, on the other hand, uses by default a flipped approach, that is, it starts from the approach of young people who have the intelligence to go, as Don Bosco said, but in reality, to use our terms, a practical dimension is enough, in the end there is no difference, right?"

"The face of the primary school children when, after an hour and a half with Comau, wait for their tablet, press "send" and the robot moves. That expression and the screams they make, I have it engraved in my head because it is the discovery of a world, that is, it is exactly what we wanted to achieve with Comau at the beginning and we achieved it, it was 'I am throwing you into a new world, and this new world is fascinating, this world is safe and this world is the future, okay?' It is not "dirty industry" that starves, kills, pollutes, it is, 'Industry is this stuff here, it is this world here that we are making you discovering now, by pressing your finger on a tablet and moving a whole series of complex phenomena that you will discover, you are in primary school now, but then you will grow', and they start from this thing here that we have

created, built"; "I don't have much direct contact with the young people, but I occasionally did some direct interventions with the young people. The thing that struck me most was their gaze. I had the perception that at that moment they were learning something they needed. The most representative image is that: having kids in front of you who are actually interested in what they are learning, who are not thinking they are wasting their time. The best image is to have perceived that if there are value propositions, the kids are able to perceive them. So maybe we get the standards we traditionally teach wrong."

"Ok, I have a thing for basketball, but I think that in the end Comau acts as a playmaker, that is, Comau's role is not to make a basket, but to assist, that is, to take a step back and say 'Okay, I do not have to prove anything to anyone, I am a multinational company", but it is giving assists and these assists are, for example, organizations like us that, just as we are accompanying companies, in turn they are in some way accompanying us, therefore this is the image representing collaboration, that is the fact of being a team where we receive an assist. We are the guard and the pivot should then be the territory, because they are the ones who have to go and make the basket."

"The image that I always have in mind is the one of the location hosting us. I have the image of the Gulf of Naples that you can see from this place, by searching online the image "Villa Doria D'Angri

" you can see it. It is one of the most prestigious villas in Naples and it is owned by the university, it is located in a spectacular position. We have received compliments from everyone for intercepting this place. This place actually represents beauty: doing something beautiful in a beautiful place makes everything perfect. This is the image I would like to take back, looking at it from the perspective that a person not only can commit and do things in the best way, with dedication, but that the context can help you too, making

Figure 12. Areas of Value questions

SECTION B – AREAS OF VALUE

1. Compared to your initiative, how important have these areas of value been during its design and implementation? ***Considering 100 as the total, give a percentage weight to each of the three areas.***
   - VALUE FOR PARTICIPANTS:
   - VALUE FOR YOUR COMPANY/ORGANIZATION:
   - VALUE FOR THE TERRITORY (e.g., scientific community, municipality, citizenship, institutions, families, etc.):
2. Do you identify other areas of value? If so, which ones?

Figure 13. Percentage distribution of the value generated by the e.DO Experience on the three levels: Participants, Company, Territory

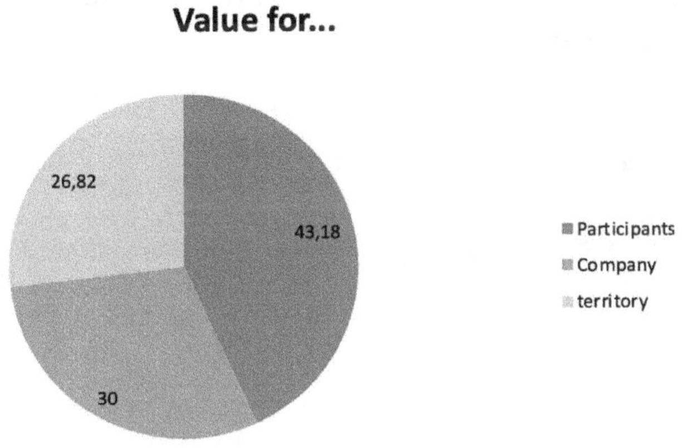

Table 1. Average responses Comau managers and partners

	MANAGER	PARTNER	OVERALL
VALUE FOR PARTICIPANTS	52.5%	32%	43.18%
VALUE FOR YOUR COMPANY/ORGANIZATION	27.5%	33%	30%
VALUE FOR THE TERRITORY	20%	35%	26.82%

everything even more beautiful. At the same time, however, you understand the beauty of the place where you are, that you can help make even more beautiful."

The evocative images selected encapsulate the key value elements described in the previous section of the chapter. Respondents make clear references to **empowering people, to live and work in the world of the future**, to the **exchange and contamination** between the different actors involved, to the **creation and development of skills**, to the **transfer of knowledge and capacitation**. Interviewees highlight **the beauty of the experience** and of **contributing to individual, social, organizational and school improvement** through **passion, engagement and motivation**.

Areas of Value

In this paragraph are considered respondents' answers with respect to section B, expressed by the question:

Figure 14. Average responses Comau managers and partners

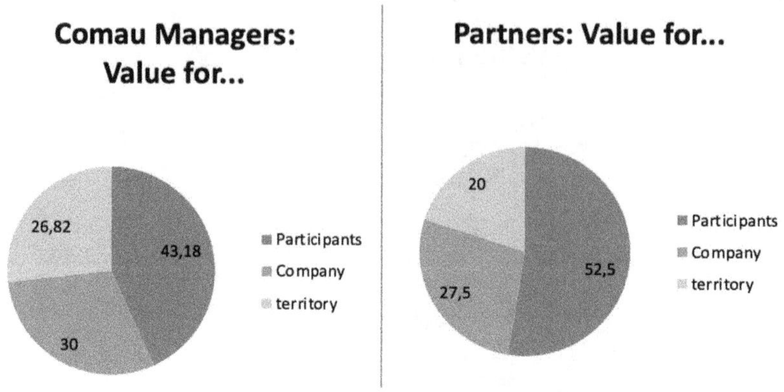

The cognitive objective is to **have a first dimensioning**, obviously without pretensions of statistical generalization, of the weight of the three areas of value according to the interviewees.

Taking an overall average of all the different responses collected, the percentage distribution shows an **imbalance in favor of Value for Participants**, which is considered overall to be the most important in a design phase. The average percentage distribution is summarized here.

- Value for Participants: 43.18%
- Value for your Company/Organization: 30%
- Value for Territory: 26.82%

This indication describes a greater importance attributed by respondents to the value generated for Participants, followed by Value for Company and Value for Territory, with the last two items being given similar importance. It is interesting to note how, by breaking down the responses according to the company they belong to (Comau vs. partner companies), **the average change significantly**. The following table and graph show the average responses of Comau managers compared with those of its partners:

There are several considerations to make: on the one hand, differences between the two groups can be observed; on the other, inter-group attributions with respect to the three levels of value.

Managers tend to explain the importance of the initiative primarily in terms of Value for Participants (52.5%), identified as the main receivers of value, while attributing less weight to the level of Value for Company (27.5%)

and Territory (20%). The importance attributed to participants is even higher than the sum of the other two areas. **Partners,** on the other hand, **tend to distribute importance in a more equitable and uniform manner**, in which the participants represent, albeit by only one percentage point, the area of value considered less important (32%), followed by Value for Company (33%), identifying the Value for Territory as the most significant aspect (35%).

Note how, although the percentages are very distinct between the two groups, the responses tend to be inverse: the difference in percentage between the two groups is 20% for the Value for Participants area (the most important for managers and the less one for partners), while that for the Territory is 15% (the most important for partners and the less one for managers). In contrast, the attribution of importance assigned to Value for Company is quite similar.

Obviously, the quantitative reading is not the object of interest in this chapter, but this section offers some interesting insights into the perceptual

Figure 15. Pillars of Value Questions

SECTION C – PILLARS OF VALUE

Below is a list of values, divided in three areas of value. Among the proposed ones, select the values that you believe are generated by your initiative.

For those selected, explain why.

1. Values for participants

AREA of VALUE	VALUE	YES	NO	WHY (for those selected)
PARTICIPANTS	Skills Creation			
	Values for New Citizenship (Human Values)			
	Empowerment			
	Inspiring & Orienteering for Working			

2. Do you identify other values? If so, which ones?
3. Values for your Company/Organization

AREA of VALUE	VALUE	YES	NO	WHY (for those selected)
COMPANY / ORGANIZATION	Core Business/Core Mission			
	Brand (Reputation/Marketing)			
	Employee Branding			
	Customer Social Responsibility (formal obligations)			
	Customer Social Responsibility (true commitment to create a better world)			
	Get external innovation potential (talents, knowhow, ideas)			
	Employees' Exposure			

4. Do you identify other values? If so, which ones?
5. Values for the territory (e.g., scientific community, municipality, citizenship, institutions, families, etc.)

AREA of VALUE	VALUE	YES	NO	WHY (for those selected)
TERRITORY (e.g., scientific community, municipality, citizenship, institutions, families, etc.)	Creation and dissemination of culture			
	Didactic Innovation			
	Contribution to the creation of new ecosystems			

6. Do you identify other values? If so, which ones?
7. If you have identified another area of value, try to indicate the corresponding values.

and meaning attribution differences found in the two groups. But then, in simple words, how can we explain these differences?

What emerges from the interviews is that managers and partners undoubtedly share the general objectives of the initiatives but, by virtue of the company's mission and degree of exposure to the territory, develop different visions: Comau is a company that deals with business, while the partners have, due to their nature, a greater orientation towards the territory. In this sense, it may be more understandable that Comau is more oriented towards the "final consumer" of the initiatives than the partners, who are much more inserted in the territorial network. As reported by many interviewees, there is a common tendency among managers to superimpose elements of value for the participants with aspects of value for the territory, as if the very fact of creating value for a participant led to creating value also for the community of which the student is part:

Interview 5: "I would give the remaining 10% to the territory, even if the 60% value for the participants actually translates into a value for the territory for what we were saying before, that is, the capacitation that concerns the students is then transmitted to a more capable and more developed territory, to a more capable community, I would say at this point ".

Interview 2P: "I see participants and territory as very interconnected, yes, because the participant in fact can never be, how to say, extrapolated from a territory, whether it is the scientific community, whether it is the municipality."

Interview 4P: "Great value can be given to the territory, because in that case the experiences that they make, now the children come to my mind, it succeeds however to sensitize also the parents, and in that case, you really have the closing of the circle, because also the territory is gratified if its students, whether they are younger or elder ones, succeed in training activities of high level."

Interview 7: "Following the instinct, maybe of that 50% that remains I would give a little more weight to the participants rather than to the territory, obviously. But it is because I cannot see the effects on the territory as a single trainer, as a single person, there is the need of some agency that takes care of it and they are taking care of it making analysis, and I think they can confirm that there is a big reflection on the territory."

Table 2. Summary synopsis of the elements collected in Section C of the qualitative interviews

LEVEL OF VALUE	AREA OF VALUE	YES	NO	WHY?
Value for Participants	Skill Creation	X		"To create STEM skills, develop school knowledge, lay a foundation in robotics and teach how to work in Industry 4.0 and the world of tomorrow."
	Values for New Citizenship	X		"You create collaborative contexts in which you can learn to work with others and with people."
	Empowerment	X		"These experiences allow you to understand your strengths and develop your skills."
	Inspiration and Orienteering for Working	X		"Many students, at the end of the class, have come up to me and asked me what they need to study to do my job."
Value for Company	Core Mission/ Core Business	X		"Superficially it is the first benefit to the company, a return in commercial and business terms."
	Brand	X		"It is a big advantage in terms of branding, both in terms of making the company more well-known and shaping its image."
	Employee Branding	X		"A child who comes to our workshops will remember us, if he or she wants to work in our field, we will be the first people they will remember."
	Formal CSR	X		"All businesses have a duty, including regulations, to contribute to society."
	CSR (True Commitment)	X		"There is a business and brand component, but the big commitment is giving back to the population."
	Open Innovation	X		"So many stakeholders come into play, from whom you can gather ideas, insights for improvement and even intercepting talent. That is the epitome of our agile approach."
	Employee's Exposure	X		"An important value is also for workers, who feel a lot of recognition, and develop their potential. This also triggers internal organizational changes that can be ridden."
Value for Territory	Creation and Dissemination of Culture	X		"The key to success is in the "train the trainer" model, in making the territory autonomous, in providing the means and know-how to help the digital and innovation culture spreading almost autonomously."
	Educational Innovation	X		"Schools must innovate to respond to social changes of which DT is just one part; e.DO offers a new educational model that is active, experiential and fun to learn from."
	Contribution to the Creation of Skills Required by the World of Work	X		"Even just developing skills that the participant can act in the territory is a very important source of value"
	Contribution to the Creation of New Ecosystems	X		"It is also important to say that this whole value is nurtured and increased through the ecosystem."

Interview 6P: "The percentage is distributed equally, 33, 33 and 33. Obviously, being a company there is a business interest, then in reality the percentage is distributed equally because there could not be one level without the other. If I do not have a territory that responds to me with participants that don't have the need, or that I cannot intercept, we could not do what we do. So, it is 33, 33, 33."

This section leaves **several insights that will be further explored through the Survey**; while managers tend to attribute considerable importance to participants, partners distribute it more evenly. We hypothesize that we will find differences in the Survey from this perspective, particularly with respect to the dimensions that constitute the level of Value for Participants.

Pillars of Value

In this last paragraph, is presented a **summary synopsis of the main elements collected** during the interviews through the questions contained in Section C of the interview outline.

On the basis of the responses collected, **a summary is presented of what emerged** in relation to the individual areas of value identified.

CONCLUSION

The interviews produced a substantial amount of material, through which it was possible to **confirm the theoretical hypotheses** developed in the previous chapters. The *e.DO Experience* **generates value on the three levels we have identified**, namely Participants, Company and Territory.

The **Value for Participants** is described mainly in terms of Creation of Skills, Empowerment and Inspiration & Orienteering for working. The aspect of value creation in terms of New Citizenship seems less evident, with the exception of the "Deal with Human and Technology" indicators, which have instead been very thoroughly examined. The Value for Participants is therefore understood in a generative perspective, oriented towards capacitation, to "make capable" to live and work in a 4.0 world, therefore, to make "digital citizens" and "community". The value that appeared to be most important is that of "Skill Creation".

The **Value for Company** is described in terms of commercial and business development, but also and above all in relation to the Brand, to market positioning, to the development of a company reputation based on social development. The value generated is also expressed in terms of the development of "company citizenship", as the staff of the companies (Comau or partners) feel a greater recognition and sense of belonging, as well as the possibility of taking part in cultural and organizational changes.

The **Value for Territory** is expressed as a function of innovation, educational and otherwise. The vehicle for value generation is represented by the creation of ecosystems and the dissemination of a culture of robotics, development and innovation. The interviewees tend to focus particularly on the value generated in terms of educational innovation, methodologies and classroom dynamics. It is precisely through the ecosystems generated and the dissemination of culture that these convey that it is possible to achieve school development goals and to create continuity in the initiative. The aspects considered most useful for the creation of social value are identified in the creation of an alliance with schools and teachers, as well as in the "train the trainer" model.

Based on the interviews evidence, authors are now able to **develop the Survey** to be administrated to a larger sample of Comau managers and partners, in order to **measure the weight** of each area of value and its respective indicators in the perception of those who designed and delivered the Comau *e.DO Experience* (as expressed in goal number 4 of our research*)*. It can therefore be assumed that the Survey will confirm and allow to size up the different areas of value, as well as the usefulness of the *e.DO Experience* in this sense. It can also be imagined that we will find differences compared to the perceptions of managers and partners with respect to the Value for Participants level, that we will get overall positive responses and that they will highlight the usefulness of the *e.DO Experience* in generating value. It is also predictable to find lower scores related to the area of Value for Participants "Value for New Citizenship", among all the less explicit one.

REFERENCES

Fregnan, E., Ivaldi, S., & Scaratti, G. (2020). HRM 4.0 and New Managerial Competences Profile: The COMAU Case. *Frontiers in Psychology*, 11.

Pinto, D., Scaratti, G., & Fregnan, E. (2021). *Learning Agility 4.0: ecosistemi e trasformazione culturale*. Pearson.

Porter, M., & Kramer, M. R. (2011). Creare valore condiviso. *Harvard Business Review Italia*, *1*(2), 68–84.

Porter, M. E., & Kramer, M. R. (2019). Creating shared value. In *Managing sustainable business* (pp. 323–346). Springer. doi:10.1007/978-94-024-1144-7_16

Richards, L., & Morse, J. M. (2013). *Readme first for a user's guide to qualitative methods*. Sage.

Chapter 7
Educational Robotics for Value Creation:
The Survey

ABSTRACT

The authors wanted to measure the perception of usefulness and the weight attributed by the promoters of the initiative (Comau managers and partners) to the Comau e.DO Experience in terms of value creation. Overall, the survey elicits a widespread perception of the usefulness of the e.DO Experience in creating value for participants, for the territory and for the company, in all their various dimensions. The greatest weight on value for participants was attributed to the "skills creation" dimension (37%), for the company to the "brand" dimension (20.13%) and for the territory to the area of educational innovation (33.02%). The survey did, however, show significant weight for all the dimensions involved, supporting the initial hypothesis that the e.DO Experience is a means of inspiring young people towards professional paths, for empowerment, the development of responsible citizenship, opening companies towards innovation, authentic corporate responsibility, and the creation of ecosystems with the territory.

DOI: 10.4018/978-1-7998-8653-2.ch007

INTRODUCTION

In the course of this chapter, authors will proceed presenting the in-depth quantitative analysis they developed based on the previous phases of the research (Richards & Morse 2013). In particular, the characteristics and types of respondents, the instrument used, the procedures followed and the main results that emerged, will be presented.

The starting hypothesis is that **there are some elements that have more impact than others in terms of Value Creation for the Participants, for the Company and for the Territory**. It is also argued that, as highlighted during the interviews, the perceptions of Comau managers and partners are often aligned, but with peculiarities linked to the belonging to one of the groups.

In fact, the analysis of the interviews presented in the previous chapter showed how Comau managers and partners share the logic of design, development and implementation of the *e.DO Experience*, attributing very aligned goals and objectives to it. However, it has been possible to note how the managers tended to attribute considerable importance to the initiative in relation to the Participants, while the partners, with a strong vision straddling the territorial and business dimensions, manifested an importance equally distributed over the three areas of value.This aspect leads the authors to hypothesize that in these two groups they **will not find particularly significant differences, except with respect to the value area concerning the Participants**. In fact, it is possible to foresee those one or more dimensions of the latter present different attributions.

PARTICIPANTS AND PROCEDURE

124 respondents took part in the survey, of which **45** are Comau managers, while **79** are members of partner companies. Participants perform different job functions, the most frequently Job Titles used by respondents were: "Teacher", "Manager"; "Consultant"; "Trainer"; "Facilitator"; "Business Development" and "Marketing Manager". Respondents have directly taken part in the delivery of Educational Robotics initiatives related to *e.DO Experience* Project and/or have taken care of the organizational aspects and design, spending on projects of different nature and in which have taken part different numbers of students; some initiatives have seen a number of participants in the order of ten, others of hundreds and others of thousands.

Respondents have managed/participated in educational robotics initiatives aimed at diverse targets:

- 37% of respondents have developed initiatives aimed at elementary school students
- 41% at middle school students
- 69% at high school students
- 32.3% at vocational school students
- 37% at university students
- 24% and 22% have addressed their initiatives to workers and executives respectively.

Participants were asked to **complete an online survey** consisting of a total of **20 questions**, organized in **four main sections**:

1. Master Data;
2. Value to Participants: Skill Creation; New Citizenship; Empowerment; Inspiration & Orienteering;
3. Value for Company;
4. Value for Territory.

At first, the questionnaire aimed to **collect preliminary personal and frame information**, related to name and professional role (1 item), organization of belonging (1 item), educational robotics initiative promoted and/or managed (1 item), number of participants involved (1 item) and type of recipients (7 items).

The questionnaire first investigated the **Value to Participants** attributed through cumulative percentage scoring to 4 categories: Skill Creation, Value for New Citizenship, Empowerment, and Inspiration and Orienteering for Working. Specifically, participants received the following instructions: “Consider 100 as the total value of the Comau Academy educational robotics initiative that you helped to promote and manage. As a percentage, how much do you think the following components impact the perceived Value for Participants? (indicate number of responses, total must equal 100):”

Subsequently, the questionnaire revealed the following dimensions related to the “Value for Participants” area, investigated through 4 measurement scales: Skill Creation (5 items, alpha = .67; Likert at 7 points); Value for New Citizenship (13 items, alpha = .94 Likert at 7 points); Empowerment (5 items, alpha = .86; Likert at 7 points); Inspiration & Orienteering (5 items, alpha = .79; Likert at 7 points).

For each of the following scales related to the Value for Participants, the questionnaire measured through 4 single items (Likert 5 points), the perception of the participant about the fact that the initiative had produced value for that specific area. The four single items were worded as follows, "In your opinion, did the people who participate in the educational robotics initiative that you helped to promote and manage, perceive value in terms of "SKILL CREATION"/"NEW CI TIZENSHIP"/"EMPOWERMENT"/"INSPIRATION & ORIENTEERING"?"

In addition, the questionnaire also included an item that delved into the "Working in the future skills" competency. Specifically, participants had to put in order by degree of solicitation the different capabilities corresponding to the above-mentioned competency.

Next, the questionnaire captured the dimension related to **Value for Company**, first by exploring the respondent's perception of utility (single item, 5-point Likert) and then through a measurement scale (5 items, alpha = .89; 7-point Likert).

The final phase of the questionnaire explored the area **Value for Territory**, first by exploring the respondent's perception of usefulness (single item, 5-point Likert) and then through the assignment of a cumulative percentage score to the 4 categories "Creation and dissemination of culture", "Educational innovation", "Contribution to the Development of Skills required by the World of Work" and "Contribution to the Creation of New Ecosystems". Specifically, participants received the following instructions: "Consider 100 as the total value of the Comau Academy educational robotics initiative that you helped to promote and manage. As a percentage, how much do you think the following components impact value sharing with the local area? (Indicate number of responses, total must equal 100)."

DATA COLLECTION AND PROCESSING

The results will be presented in line with the areas that structured the questionnaire, in order to offer the main descriptive and inferential statistics to support the produced output.

Value for Participants

This level of the value generated by Educational Robotics initiatives was explored through the use of a single item (5-point Likert scale) and four scales referring to the areas of values: "Skill Creation"; "New Citizenship"; "Empowerment"; "Inspiration & Orienteering".

With the first item are measured, in percentages, the weight that different respondents give to the four areas of the "Value for Participants" level. This is the item, as the respondents read it:

Consider 100 as the total value of the Comau Academy educational robotics initiative that you helped to promote and manage. As a percentage, how much do you think the following components impact the perceived value of the participants? (indicate number of responses, total must equal 100):

- Skill Creation (creation of technical, curricular, robotic and relational skills that will enable tomorrow's world of work) (Fregnan et al., 2020)
- Value for New Citizenship (being responsible citizens, able to step out of one's comfort zone, respectful of others and rules)
- Empowerment (becoming aware of one's abilities, being resourceful and proactive)
- Inspiration & Orienteering for Working (generating passion, providing professional role models through engaging experiences relating to technologies)

The descriptive analysis allowed us to define the average score, in percentage, for each of the following areas: a summary table is given below in which the mean and standard deviation of each item are reported.

From a purely descriptive standpoint, the area indicated as most relevant in terms of Value Creation for the Participant is "Skill Creation" (35.2%), followed by "Inspiration & Orienteering for Working" (27.6%), Empowerment (22.9) and finally "Value for new Citizenship" (27.6%).

Descriptive percentages of the distribution are briefly shown in Figure 1.

Table 1. Average score in percentage for each area related to the level "Value for Participants"

Sub-dimension:	Media	Std. deviation
SKILL CREATION	35.2%	16.82
VALUE FOR NEW CITIZENSHIP	14.3%	8.11
EMPOWERMENT	22.9%	11.41
INSPIRATION & ORIENTEERING	27.6%	12.13

Figure 1. Percentages of the distribution for the areas of value related to the level "Value for Participants"

1. Name and Role
2. Short description of the educational robotics initiative (name, participants' description, place, time, goals and program, results)
3. What did your company/organization and you deal with within the educational robotics initiative?

SECTION A - OVERALL

1. What value do you think the proposed initiative has for the PARTICIPANTS (what was it for, what did they bring home)? Tell an episode to exemplify the value you perceive.
2. What value do you think the proposed initiative has for the COMPANY/ORGANIZATION (what was it for, what did they bring home)? Tell an episode to exemplify the value you perceive.
3. What value do you think the proposed initiative has for the TERRITORY (e.g., scientific community, municipality, citizenship, institutions, families, etc.) (what was it for, what did they bring home)? Tell an episode to exemplify the value you perceive.
4. What is the image (concept) that best represents the value / values you have identified?

SECTION B – AREAS OF VALUE

1. Compared to your initiative, how important have these areas of value been during its design and implementation? *Considering 100 as the total, give a percentage weight to each of the three areas.*

- VALUE FOR PARTICIPANTS:
- VALUE FOR YOUR COMPANY/ORGANIZATION:
- VALUE FOR THE TERRITORY (e.g., scientific community, municipality, citizenship, institutions, families, etc.):

2. Do you identify other areas of value? If so, which ones?

SECTION C – PILLARS OF VALUE

Below is a list of values, divided in three areas of value. Among the proposed ones, select the values that you believe are generated by your initiative.

For those selected, explain why.

1. Values for participants

AREA of VALUE	VALUE	YES	NO	WHY (for those selected)
PARTICIPANTS	Skills Creation			
	Values for New Citizenship (Human Values)			
	Empowerment			
	Inspiring & Orienteering for Working			

2. Do you identify other values? If so, which ones?

3. Values for your Company/Organization

AREA of VALUE	VALUE	YES	NO	WHY (for those selected)
COMPANY / ORGANIZATION	Core Business/Core Mission			
	Brand (Reputation/Marketing)			
	Employee Branding			
	Customer Social Responsibility (formal obligations)			
	Customer Social Responsibility (true commitment to create a better world)			
	Get external innovation potential (talents, knowhow, ideas)			
	Employees' Exposure			

4. Do you identify other values? If so, which ones?

5. Values for the territory (e.g., scientific community, municipality, citizenship, institutions, families, etc.)

AREA of VALUE	VALUE	YES	NO	WHY (for those selected)
TERRITORY (e.g., scientific community, municipality, citizenship, institutions, families, etc.)	Creation and dissemination of culture			
	Didactic Innovation			
	Contribution to the creation of new ecosystems			

6. Do you identify other values? If so, which ones?

7. If you have identified another area of value, try to indicate the corresponding values.

Using the descriptive statistics above, it is possible to size the four sub-dimensions of "Value for Participants" according to a valid average percentage for the sample we used.

In the next step, typical of inferential statistics, authors extended the survey to the wider population, with the aim of generalizing these results. The comparison between averages made it possible to highlight that the difference between the averages was statistically significant and, consequently, it is possible to generalize within the population the fact that the item "Skill Creation" has a greater weight than "New Citizenship" (F (1, 123) = 109.45, $p \leq .001$), "Empowerment" (F (1, 123) = 30.92, $p < .001$) and "Inspiration & Orienteering" (F (1, 123) = 10.95, p = .001). Authors then wanted to see if being a Manager at Comau or a Partner had any statistically significant difference with respect to the areas mentioned above; inferential statistics show that for the items "Skill Creation" (t (122) = 1.38, p = .169), "New Citizenship" (t (122) = .761, p = .448) and "Empowerment" (t (122) = .450, p = .654) do not emerge statistically significant differences related to the organization the respondents belong to (1 = Comau; 2 = Partner Company), while it can be observed that Comau managers attribute greater weight to the item "Inspiration & Orienteering" (t (122) = 2.02, p = .046) than Partners. It can consequently be inferred that the items "Skill Creation", "New Citizenship" and "Empowerment" have the same weight for both Comau Managers and Partners, while the item "Inspiration and Orienteering for Working" is considered more relevant by Comau Managers (30.44% for Managers vs. 25.92% for Partners).

In the next step, authors went to size the different areas of the "Value for Participants" level.

As had already been hypothesized following the literature review and the previous qualitative phase, **Skills Creation occupies a central position in the creation of Value for Participants**. What strikes one as very positive is the evidence linked to the other three areas of this level of value: in fact, it can be seen that the respondents consider the initiative valid not only for **indicating possible training and professional trajectories** to the participants, but at the same time it represents a means of **becoming aware of one's own strengths** and **developing one's own abilities**, as well as - and this is an element that strikes one particularly - through e.DO it is possible to **become better citizens**, attentive to others and to one's own actions in society.

Skill Creation

Initially, authors asked the respondent via a single item (5-step Likert Scale) whether the participants had perceived value in terms of "Skill Creation". The question was phrased as follows:

In your opinion, did the people who participated in the educational robotics initiative that you helped to promote and manage perceive value in terms of "Skill Creation" (creation of technical, curricular, robotic and relational skills that will enable tomorrow's world of work)?

Table 2 shows descriptive statistics related to this area of value.

It can therefore be observed here, from a purely descriptive point of view valid just within our sample, that most of the respondents believe that the Participants created value in terms of Skill Creation (56.5% absolutely yes, 40.3% probably yes). In this case, the answer does not appear to be influenced by the respondent's organization ($t(122) = 1.457, p = .148$). Specifically, as shown in the table above, the question in object included 5 possible answers indicative of the presence or absence of a usefulness of the *e.DO Experience* in terms of Skill Creation:

- Absolutely yes
- Probably yes
- Neither yes nor no
- Probably not
- Absolutely not

Precisely, the responding population rated the usefulness of *e.DO Experience* initiatives in terms of Skill Creation as follows:

- 56.5% believes that the *e.DO Experience* is absolutely useful in generating value in terms of Skill Creation

Table 2. Descriptive statistics related to the "Skill Creation" item

Skill Creation Value Perception	Frequency	Percentage
Absolutely yes	70	56.5
Probably yes	50	40.3
Neither yes nor no	3	2.4
Probably not	1	.8
Absolutely not	0	0

Figure 2. Percentages of the distribution of value in the item "Skills Creation"

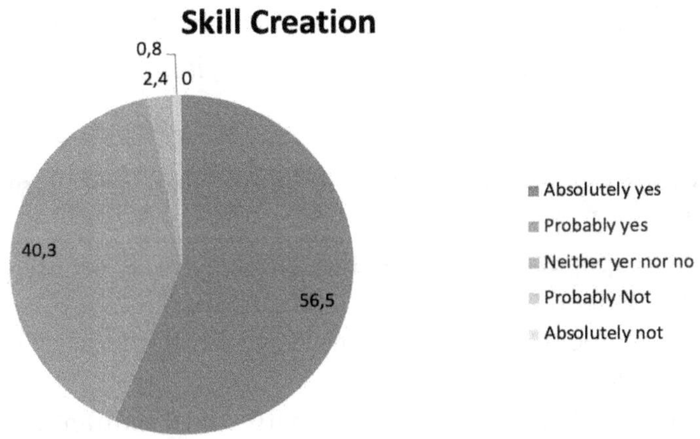

- 40.3% believes that the *e.DO Experience* is likely to be useful in generating value in terms of Skill Creation
- 2.4% believes that the *e.DO Experience* is neither useful nor unhelpful in generating value in terms of Skill Creation
- 0.8% believes that the *e.DO Experience* is probably not useful for generating value in terms of Skill Creation
- No respondents believe that the *e.DO Experience* is completely unhelpful in generating value in terms of Skill Creation

Next, authors sized the usefulness of the initiative in terms of "Skill Creation" through a scale (5 items, alpha = .67; 7-point Likert). The different Items composing the scale allowed then to size the respondents' perceived usefulness in terms of "Robotic Skills", "School Subject Skills", "Professional Hard Skills", "Soft Skills" and "Working in the future Skills". Here are reported descriptive statistics for each one, considering that the respondent, for each question, had seven answer choices to indicate the degree of usefulness of the initiative with respect to each dimension/value indicator:

- Very useful
- Useful
- Quite useful
- Indifferent
- Not very useful
- Not useful
- Totally useless

Robotic Skills

With respect to the question "How useful do you think the *e.DO Experience* initiative was for the creation of Robotic Skills?", it is possible to summarize the distribution of responses as follows:

- 43.5% of respondents felt that the initiative was very helpful in terms of developing Robotic Skills
- 32.3% of respondents felt that the initiative was useful in terms of developing Robotic Skills
- 21% of respondents felt that the initiative was quite useful in terms of developing Robotic Skills
- 2.4% of respondents felt that the initiative was indifferent in terms of developing Robotic Skills
- 0% of respondents felt that the initiative was of little use in terms of developing Robotic Skills
- 0.8% of respondents felt that the initiative was not useful in terms of developing Robotic Skills
- 0% of respondents felt that the initiative was totally unnecessary in terms of developing Robotic Skills

Therefore, it can be clearly observed that most of the participants believe that the Educational Robotics project was useful for the development of "Robotic Skills", only one participant believes that it was not useful.

School Subject Skills

With respect to the question "How useful do you think the *e.DO Experience* initiative has been for the creation of School Subject Skills?", it is possible to summarize the distribution of responses as follows:

- 18.5% of respondents felt that the initiative was very helpful in terms of developing School Subject Skills
- 41.1% of respondents felt that the initiative was helpful in terms of developing School Subject Skills
- 28.2% of respondents felt that the initiative was quite helpful in terms of developing School Subject Skills
- 6.5% of respondents felt that the initiative was indifferent in terms of developing School Subject Skills

Figure 3. Distribution of responses related to the "School Subject Skills" item

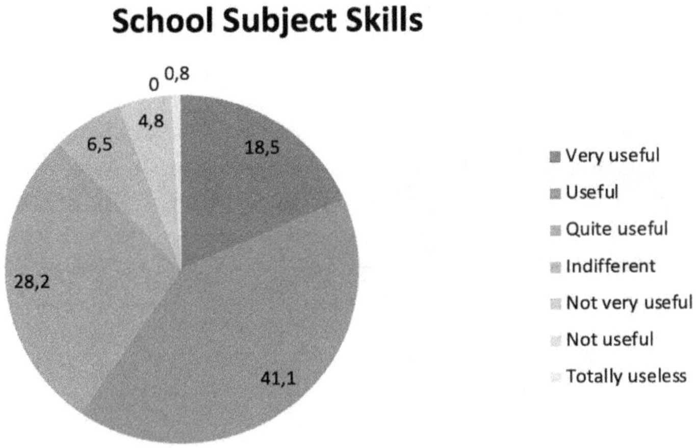

- 4.8% of respondents felt that the initiative was of little use in terms of developing School Subject Skills
- 0% of respondents felt that the initiative was not helpful in terms of developing School Subject Skills
- 0.8% of respondents felt that the initiative was totally unnecessary in terms of developing School Subject Skills

Consequently, it can be clearly seen that the majority of respondents felt that the Educational Robotics project was useful in developing "School Subject skills". In fact, a slightly lower percentage than 90% of respondents has attributed the initiative a level of usefulness.

Professional Hard Skills

With respect to the question "How useful do you think the *e.DO Experience* initiative was for the creation of Professional Hard Skills?", it is possible to summarize the distribution of responses as follows:

- 16.9% of respondents felt that the initiative was very helpful in terms of developing Professional Hard Skills
- 38.7% of respondents felt that the initiative was helpful in terms of developing Professional Hard Skills
- 29% of respondents felt that the initiative was quite useful in terms of developing Professional Hard Skills

Figure 4. Distribution of responses related to the "Professional Hard Skills" item

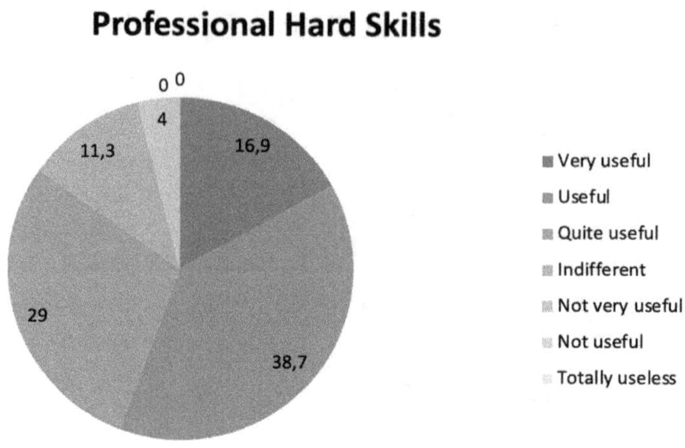

- 11.3% of respondents felt that the initiative was indifferent in terms of developing Professional Hard Skills
- 4% of respondents felt that the initiative was of little use in terms of developing Professional Hard Skills
- 0% of respondents felt that the initiative was not helpful in terms of developing Professional Hard Skills
- 0% of respondents felt that the initiative was totally unnecessary in terms of developing Professional Hard Skills

Thus, it can be clearly seen that the majority of participants believe that the Educational Robotics project was useful in developing "Professional Hard Skills". Only 4% of respondents attributed little usefulness to the experience, while 11% considered it as indifferent.

Soft Skills

With respect to the question "How useful do you think the *e.DO Experience* initiative was for the creation of Soft Skills?", it is possible to summarize the distribution of responses as follows:

- 33.1% of respondents felt that the initiative was very helpful in terms of developing Soft Skills
- 35.5% of respondents felt that the initiative was helpful in terms of developing Soft Skills

Figure 5. Distribution of responses related to the "Soft Skills" item

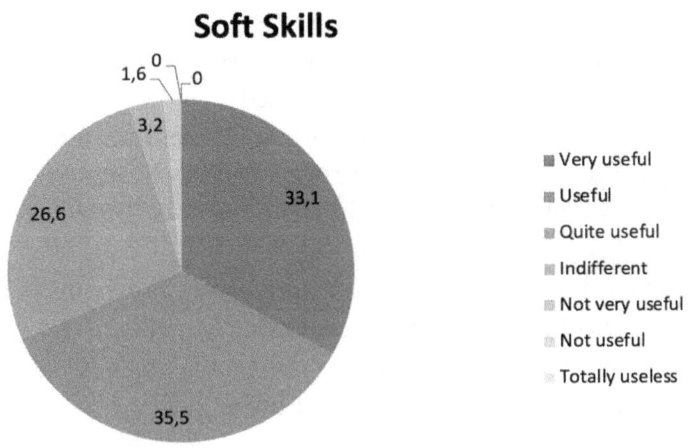

- 26.6% of respondents felt that the initiative was quite useful in terms of developing Soft Skills
- 3.2% of respondents felt that the initiative was indifferent in terms of developing Soft Skills
- 1.6% of respondents felt the initiative was of little use in terms of developing Soft Skills
- 0% of respondents felt that the initiative was not useful in terms of developing Soft Skills
- 0% of respondents felt that the initiative was totally unnecessary in terms of developing Soft Skills

Therefore, you can clearly see that most of the participants believe that the Educational Robotics project was useful for the development of "Soft Skills"; only 6 respondents considered the initiative as not very useful or as indifferent.

Working in the Future Skills

With respect to the question "How useful do you think the *e.DO Experience* initiative was for the creation of Working in the Future Skills?", it is possible to summarize the distribution of responses as follows:

- 30.6% of respondents felt that the initiative was very helpful in terms of developing Working in the Future Skills

- 39.5% of respondents felt that the initiative was helpful in terms of developing Working in the Future Skills
- 27.4% of respondents felt that the initiative was quite helpful in terms of developing Working in the Future Skills
- 2.4% of respondents felt that the initiative was indifferent in terms of developing Working in the Future Skills
- 0% of respondents felt that the initiative was of little use in terms of developing Working in the Future Skills
- 0% of respondents felt that the initiative was not helpful in terms of developing Working in the Future Skills
- 0% of respondents felt that the initiative was totally unnecessary in terms of developing Working in the Future Skills

Consequently, it can be clearly seen that most participants felt that the **Educational Robotics projecct was useful in developing "Working in the Future Skills"**. In fact, no participant answered that the initiative was unhelpful or totally unnecessary.

The "Skill Creation" section ended with an in-depth study of one of the different classes of skills investigated, renamed Working in the Future Skills, and composed of 8 capabilities. Participants were asked to sort the different capabilities according to how much they were solicited by the Educational Robotics project (1 the most solicited, 8 the less solicited). The results allowed for the ordering of the capabilities based on the following ranking:

Figure 6. Distribution of responses related to the "Working in the Future Skills" item

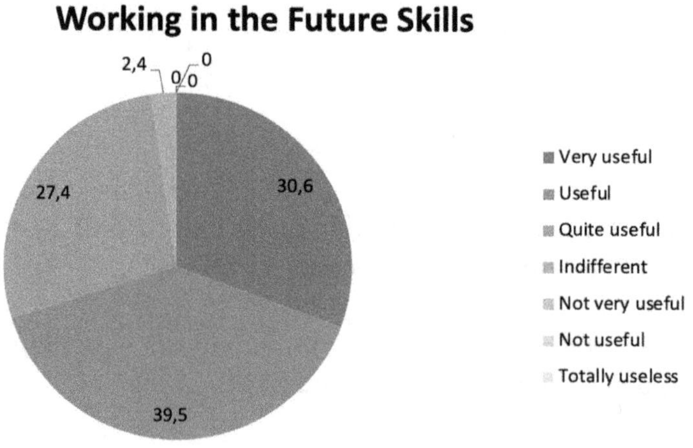

1. Deal with Technology
2. Innovation
3. Collaboration
4. Engagement
5. Deal with Human
6. Agility
7. Interdisciplinarity
8. Critical Thinking

From the point of view of respondents, **the capability "Deal with Technology" appears to be the most stimulated** by the Educational Robotics experiences developed around the e.DO Robot, followed in order by "Innovation", "Collaboration", "Engagement", "Deal with Human", "Agility", "Interdisciplinarity" and, in last place, "Critical Thinking".

Value for New Citizenship

Authors firstly asked the respondent via a single item (5-Step Likert Scale) whether participants had perceived value in terms of "New Citizenship." The question was phrased as follows:

In your opinion, did the people who participated in the training initiative you helped to promote and manage perceive value in terms of "VALUE FOR NEW CITIZENSHIP" (Being responsible citizens, able to step out of their comfort zone, respectful of others and of the rules)?

Specifically, as shown in the table below, the question in object included 5 possible answers indicative of the presence or absence of usefulness of the *e.DO Experience*, in terms of Value for new Citizenship:

- Absolutely yes
- Probably yes
- Neither yes nor no
- Probably not
- Absolutely not

Precisely, the responding population rated the usefulness of *e.DO Experience* initiatives in terms of Value for new Citizenship as follows:

- 19.4% believes that the *e.DO Experience* is absolutely useful in generating value in terms of Value for new Citizenship
- 53.2% believes that the *e.DO Experience* is likely to be useful in generating value in terms of Value for new Citizenship
- 21.8% believes that the *e.DO Experience* is neither useful nor unhelpful in generating value in terms of Value for new Citizenship
- 5.6% believes that the *e.DO Experience* is probably not useful for generating value in terms of Value for new Citizenship
- No respondents believe that the *e.DO Experience* is completely unhelpful in generating value in terms of Value for new Citizenship

Table 3 and Figure 7 shows the descriptive statistics related to this area of value.

Table 3. Descriptive statistics related to the "Value for new Citizenship" area

New CITIZENSHIP Value Perception	Frequency	Percentage
Absolutely yes	24	19.4
Probably yes	66	53.2
neither yes nor no	27	21.8
Probably not	7	5.6
Absolutely not	0	0

Figure 7. Percentages of the distribution of value in the area "Value for New Citizenship"

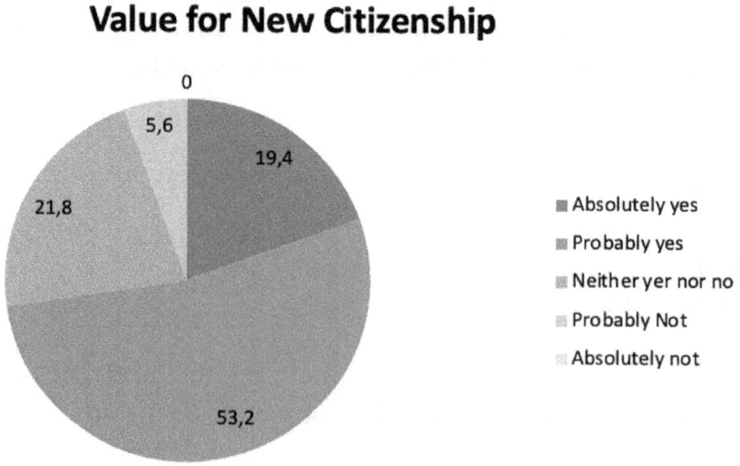

Thus, from a purely descriptive point of view and valid just within our sample, it can be seen here that a small portion of the respondents (19.4%) claims with absolute certainty that the participants created value in terms of new citizenship, while half of the respondents (53.2%) expresses a good probability that the participants perceived this value. In this case, the response does not appear to be influenced by the respondent's membership organization ($t\ (122) = .04$, $p = .968$).

Next, authors sized the usefulness of the initiative in terms of "Value for New Citizenship" through a scale (13 items, alpha = .94 7-point Likert). Of each item comprising the "Value for New Citizenship" scale, are reported descriptive statistics.

Challenging

With respect to the question "How useful do you think the *e.DO Experience* initiative was in developing the ability to Challenge oneself?", it is possible to summarize the distribution of responses as follows:

- 30.6% of respondents felt that the initiative was very helpful in terms of developing the ability to Challenge themselves
- 37.9% of respondents felt that the initiative was helpful in terms of developing the ability to Challenge themselves
- 25% of respondents felt that the initiative was quite helpful in terms of developing the ability to Challenge themselves
- 4% of respondents felt that the initiative was indifferent in terms of developing the ability to Challenge themselves
- 1.6% of respondents felt that the initiative was unhelpful in terms of the ability to Challenge themselves
- 0% of respondents felt that the initiative was not helpful in terms of developing the ability to Challenge themselves
- 0.8% of respondents felt that the initiative was totally unhelpful in terms of developing the ability to Challenge themselves

You can therefore clearly see that most of the participants believe that the Educational Robotics project has been useful for the development of the indicator "Challenge"; only 8 respondents have not attributed usefulness to the initiative.

Respecting Others

With respect to the question "How useful do you think the *e.DO Experience* initiative has been for the development of the indicator Respecting Others?", it is possible to summarize the distribution of responses as follows:

- 21% of respondents felt that the initiative was very useful in terms of developing the indicator Respecting Others
- 35.5% of respondents felt that the initiative was useful in terms of developing the indicator Respecting Others
- 25.8% of respondents felt that the initiative was quite useful in terms of developing the indicator Respecting Others
- 12.1% of respondents felt that the initiative was indifferent in terms of the development of the indicator Respecting Others
- 4.8% of respondents felt that the initiative was of little use in terms of developing the indicator Respecting Others
- 0.8% of respondents felt that the initiative was not useful in terms of developing the indicator Respecting Others
- 0% of respondents felt that the initiative was totally unnecessary in terms of developing the indicator Respecting Others

Therefore, it can be clearly observed that the majority of participants believes that the Educational Robotics project has been useful for the development of the indicator "Respecting Others"; in fact, more than 80% of respondents has attributed usefulness in this sense

Figure 8. Distribution of responses related to the "Respecting Others" item

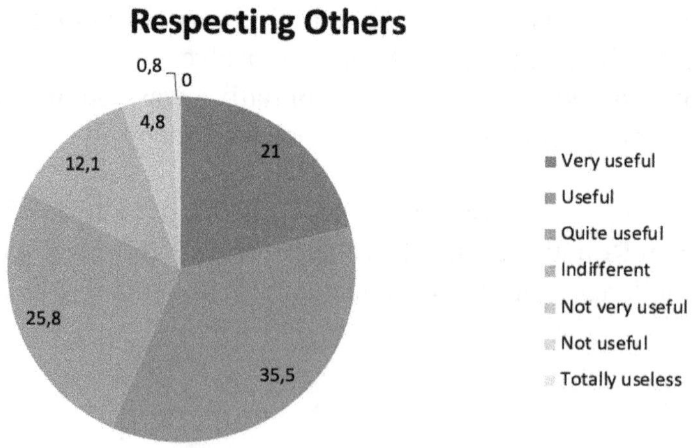

Respecting the Rules

With respect to the question "How useful do you think the *e.DO Experience* initiative has been for the development of the Respecting the Rules indicator?", it is possible to summarize the distribution of responses as follows:

- 29% of respondents felt that the initiative was very helpful in terms of developing the indicator Respecting the Rules
- 39.5% of respondents felt that the initiative was useful in terms of developing the indicator Respecting the Rules
- 18.5% of respondents felt that the initiative was quite useful in terms of developing the indicator Respecting the Rules
- 11.3% of respondents felt that the initiative was indifferent in terms of the development of the indicator Respecting the Rules
- 0.8% of respondents felt that the initiative was of little use in terms of developing the indicator Respecting the Rules
- 0.8% of respondents felt that the initiative was not useful in terms of developing the indicator Respecting the Rules
- 0% of respondents felt that the initiative was totally unnecessary in terms of developing the indicator Respecting the Rules

Thus, it can be clearly seen that the majority of participants believe that the Educational Robotics project was useful for the development of the "Respecting the Rules" indicator. In fact, a percentage of just under 90% attributed usefulness in this sense.

Figure 9. Distribution of responses related to the "Respecting the Rules" item

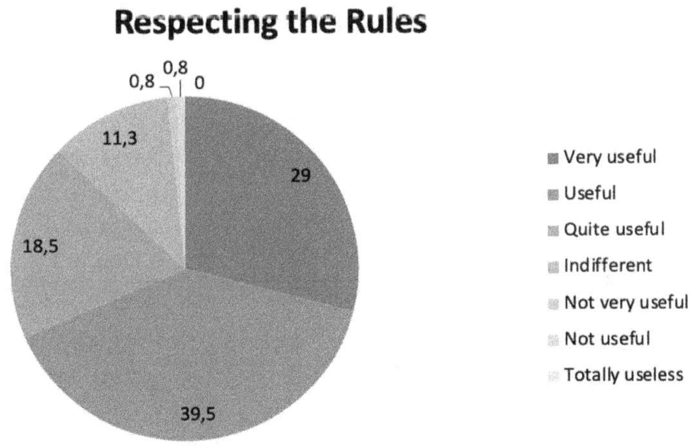

Generosity and Commitment

With respect to the question "How useful do you think the *e.DO Experience* initiative has been for the development of the Generosity and Commitment indicator?", it is possible to summarize the distribution of responses as follows:

- 21% of respondents felt that the initiative was very helpful in terms of developing the Generosity and Commitment indicator
- 35.5% of respondents felt that the initiative was helpful in terms of developing the Generosity and Commitment indicator
- 31.5% of respondents felt that the initiative was quite useful in terms of developing the Generosity and Commitment indicator
- 8.9% of respondents felt that the initiative was indifferent in terms of the development of the Generosity and Commitment indicator
- 2.4% of respondents felt that the initiative was of little use in terms of developing the Generosity and Commitment indicator
- 0.8% of respondents felt that the initiative was not helpful in terms of developing the Generosity and Commitment indicator
- 0% of respondents felt that the initiative was totally unnecessary in terms of developing the Generosity and Commitment indicator

Therefore, it can be clearly observed that the majority of participants believes that the Educational Robotics project was useful for the development of the "Generosity and Commitment" indicator; only 15 participants did not consider the initiative useful in this sense.

Figure 10. Distribution of responses related to the "Generosity and Commitment" item

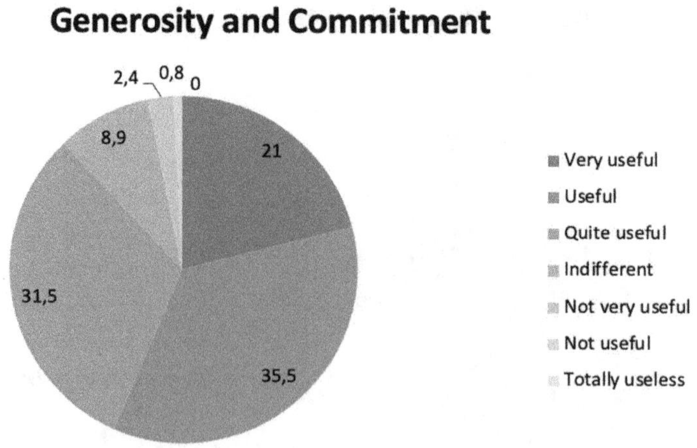

Exemplary and Consistent

With respect to the question "How useful do you think the *e.DO Experience* initiative was for the development of the indicator Being Exemplary and Consistent?", it is possible to summarize the distribution of responses as follows:

- 12.9% of respondents felt that the initiative was very helpful in terms of developing the indicator Being Exemplary and Consistent
- 25% of respondents felt that the initiative was helpful in terms of developing the indicator Being Exemplary and Consistent
- 31.5% of respondents felt that the initiative was quite useful in terms of developing the indicator Being Exemplary and Consistent
- 25% of respondents felt that the initiative was indifferent in terms of developing the indicator Being Exemplary and Consistent
- 4.8% of respondents felt that the initiative was of little use in terms of developing the indicator Being Exemplary and Consistent
- 0.8% of respondents felt that the initiative was not helpful in terms of developing the indicator Being Exemplary and Consistent
- 0% of respondents felt that the initiative was totally unnecessary in terms of developing the indicator Being Exemplary and Consistent

Thus, it can be clearly seen that the majority of respondents felt that the Educational Robotics project was helpful in developing the "Being Exemplary and Consistent" indicator. However, it can be noticed that one in

Figure 11. Distribution of responses related to the "Exemplary and Consistent" item

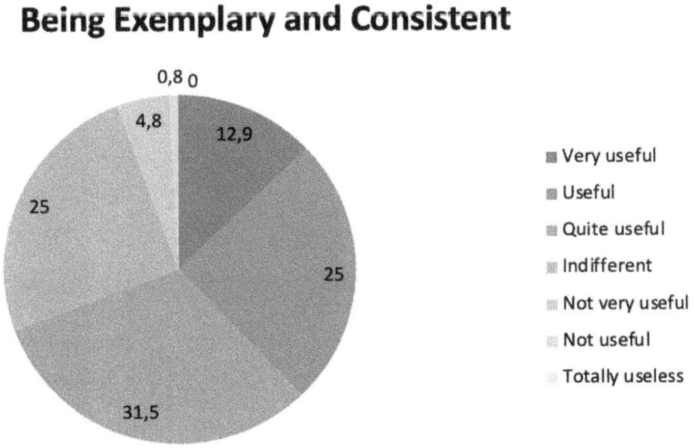

four respondents felt that the initiative was indifferent in this regard. Overall, the indicator "Being Exemplary and Consistent" is one of those that received one of the lowest scores in terms of usefulness from respondents, although almost 70% responded positively.

Keeping Promises

With respect to the question "How useful do you think the *e.DO Experience* initiative has been for the development of the Keeping the Promises Indicator?", it is possible to summarize the distribution of responses as follows:

- 15.3% of respondents felt that the initiative was very helpful in terms of developing the indicator Keeping Promises
- 21% of respondents felt that the initiative was helpful in terms of developing the indicator Keeping Promises
- 22.6% of respondents felt that the initiative was quite helpful in terms of developing the indicator Keeping Promises
- 34.7% of respondents felt that the initiative was indifferent in terms of developing the indicator Keeping Promises
- 3.2% of respondents felt that the initiative was of little use in terms of developing the indicator Keeping Promises
- 2.4% of respondents felt that the initiative was not helpful in terms of developing the indicator Keeping Promises
- 0.8% of respondents felt that the initiative was totally unnecessary in terms of developing the indicator Keeping Promises

Figure 12. Distribution of responses related to the "Keeping Promises" item

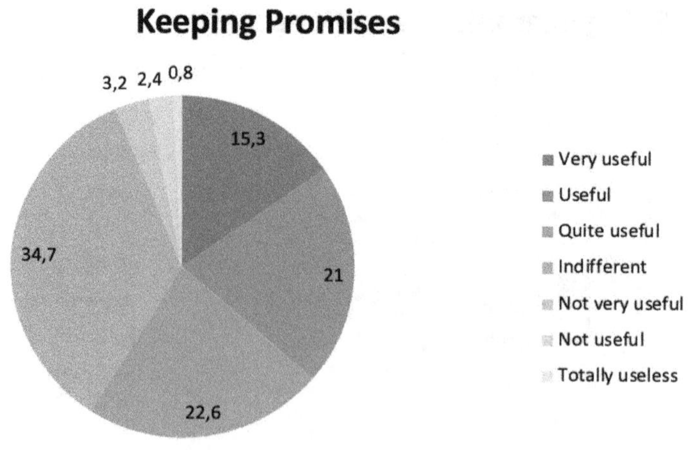

It can be seen that about 40% of respondents considered the initiative indifferent or not useful in this regard. Overall, the indicator "Keeping Promises" is one of those that received one of the lowest scores in terms of usefulness from respondents; positive responses came from less than 60% of respondents.

Measuring Results Objectively

With respect to the question "How useful do you think the *e.DO Experience* initiative has been for the development of the indicator Measuring Results Objectively?", it is possible to summarize the distribution of responses as follows:

- 31.5% of respondents felt that the initiative was very useful in terms of developing the indicator Measuring Results Objectively
- 40.3% of respondents felt that the initiative was useful in terms of developing the indicator Measuring Results Objectively
- 16.9% of respondents felt that the initiative was quite useful in terms of developing the indicator Measuring Results Objectively
- 10.5% of respondents felt that the initiative was indifferent in terms of developing the indicator Measuring Results Objectively
- 0.8% of respondents felt that the initiative was of little use in terms of developing the indicator Measuring Results Objectively

Figure 13. Distribution of responses related to the "Measuring the Results Objectively" item

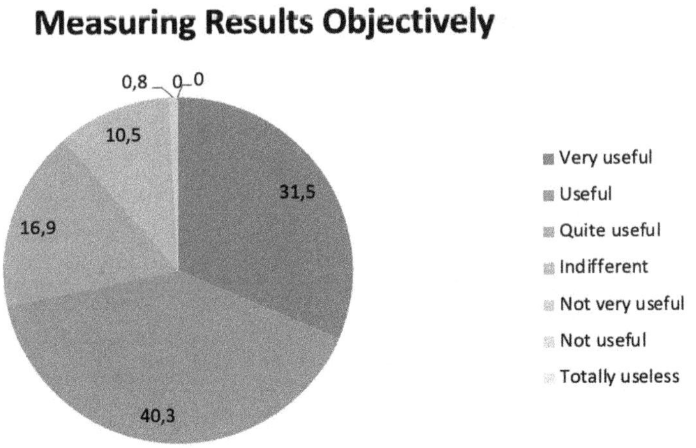

- 0% of respondents felt that the initiative was not useful in terms of developing the indicator Measuring Results Objectively
- 0.8% of respondents felt that the initiative was totally unnecessary in terms of developing the indicator Measuring Results Objectively

Thus, it can be clearly seen that most participants believe that the educational robotics initiative was useful for the development of the "Measuring Results Objectively" indicator. Only one participant considers it as not very useful; no one indicates an absence of usefulness, while about 10% considers it as indifferent.

Responsibility and Reciprocity

With respect to the question "How useful do you think the *e.DO Experience* initiative was for the development of the Responsibility and Reciprocity indicator?", it is possible to summarize the distribution of responses as follows:

- 18.5% of respondents felt that the initiative was very helpful in terms of developing the Responsibility and Reciprocity indicator
- 36.3% of respondents felt that the initiative was helpful in terms of developing the Responsibility and Reciprocity indicator
- 27.4% of respondents felt that the initiative was quite useful in terms of developing the Responsibility and Reciprocity indicator

Figure 14. Distribution of responses related to the "Responsibility and Reciprocity" item

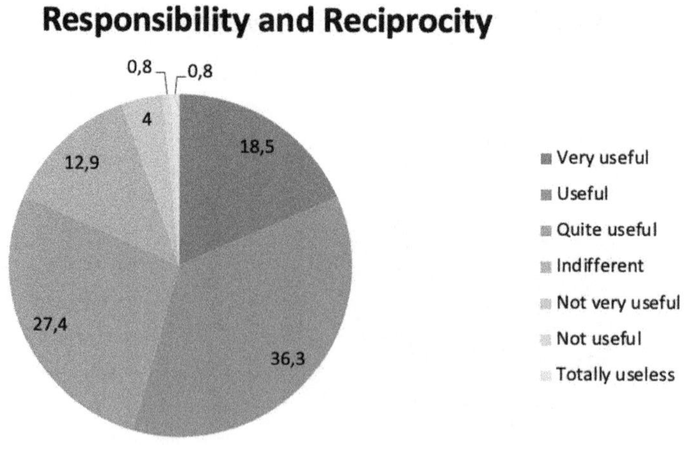

- 12.9% of respondents felt that the initiative was indifferent in terms of developing the Responsibility and Reciprocity indicator
- 4% of respondents felt that the initiative was of little use in terms of developing the Responsibility and Reciprocity indicator
- 0.8% of respondents felt that the initiative was not helpful in terms of developing the Responsibility and Reciprocity indicator
- 0.8% of respondents felt that the initiative was totally unnecessary in terms of developing the Responsibility and Reciprocity indicator

Therefore, it can be clearly seen that the majority of participants believes that the Educational Robotics project has been useful for the development of the indicator "Responsibility and Reciprocity"; in fact, more than 85% of respondents attributes a usefulness in this sense.

Equity and Justice

With respect to the question "How useful do you think the *e.DO Experience* initiative has been for the development of the Equity and Justice indicator?", it is possible to summarize the distribution of responses as follows:

- 15.3% of respondents felt that the initiative was very helpful in terms of developing the Equity and Justice indicator
- 17.7% of respondents felt that the initiative was helpful in terms of developing the Equity and Justice indicator
- 22.6% of respondents felt that the initiative was quite useful in terms of developing the Equity and Justice indicator
- 35.5% of respondents felt that the initiative was indifferent in terms of developing the Equity and Justice indicator
- 4.8% of respondents felt that the initiative was of little use in terms of developing the Equity and Justice indicator
- 4% of respondents felt that the initiative was not helpful in terms of developing the Equity and Justice indicator
- 0.8% of respondents felt that the initiative was totally unnecessary in terms of developing the Equity and Justice indicator

For this item, it can be observed that only slightly more than 50% of the respondents considered the educational robotics initiative useful for the development of the "Equity and Justice" indicator. Overall, the "Equity and

Figure 15. Distribution of responses related to the "Equity and Justice" item

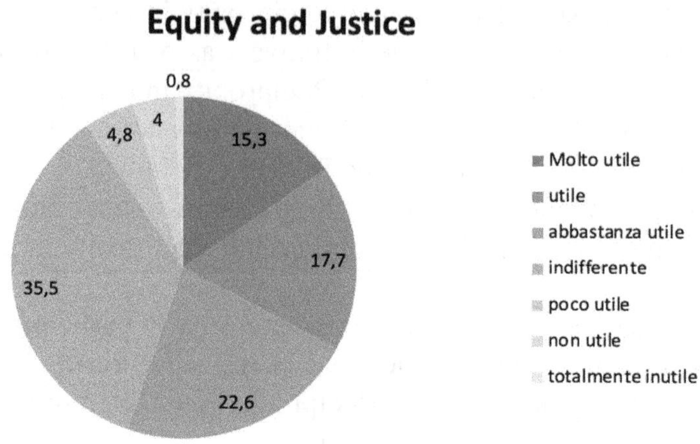

Justice" indicator is one of those that received one of the lowest scores in terms of usefulness from respondents, as positive responses came from less than 60% of respondents.

Being Transparent and Sincere

With respect to the question "How useful do you think the *e.DO Experience* initiative has been for the development of the indicator Being Transparent and Sincere?", it is possible to summarize the distribution of responses as follows:

- 15.3% of respondents felt that the initiative was very useful in terms of developing the indicator Being Transparent and Sincere
- 17.7% of respondents felt that the initiative was useful in terms of developing the indicator Being Transparent and Sincere
- 25.8% of respondents felt that the initiative was quite useful in terms of the indicator Being Transparent and Sincere
- 35.5% of respondents felt that the initiative was indifferent in terms of the development of the indicator Being Transparent and Sincere
- 4% of respondents felt that the initiative was of little use in terms of developing the indicator Being Transparent and Sincere
- 1.6% of respondents felt that the initiative was not useful in terms of developing the indicator Being Transparent and Sincere
- 0% of respondents felt that the initiative was totally unnecessary in terms of developing the indicator Being Transparent and Sincere

Figure 16. Distribution of responses related to the "Be Transparent and Sincere" item

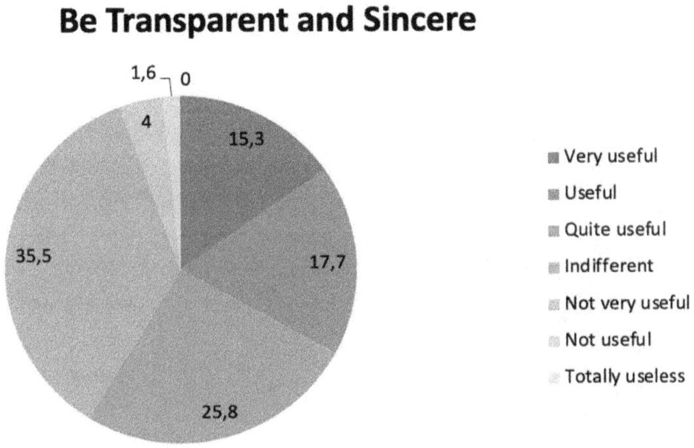

For this item, it can be observed that only a bit more than 50% of the respondents considered the educational robotics initiative useful for the development of the "Being Transparent and Sincere" indicator. Overall, the indicator "Being Transparent and Sincere" is one of those that received one of the lowest scores in terms of usefulness from respondents, in fact positive responses came from less than 60% of respondents.

Comparison and Exchange

With respect to the question "How useful do you think the *e.DO Experience* initiative has been for the development of the Comparison and Exchange indicator?", it is possible to summarize the distribution of responses as follows:

- 33.9% of respondents felt that the initiative was very useful in terms of developing the indicator Comparison and Exchange
- 37.1% of respondents believe that the initiative was useful in terms of developing the indicator Comparison and Exchange
- 21% of respondents felt that the initiative was quite useful in terms of developing the indicator Comparison and Exchange
- 7.3% of respondents felt that the initiative was indifferent in terms of developing the indicator Comparison and Exchange
- 0.8% of respondents felt that the initiative was of little use in terms of developing the indicator Comparison and Exchange
- 0% of respondents felt that the initiative was not useful in terms of developing the indicator Comparison and Exchange

Figure 17. Distribution of responses related to the "Comparison and Exchange" item

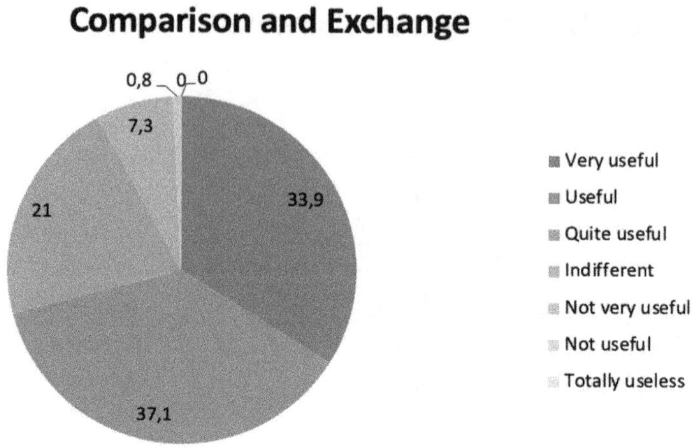

- 0% of the respondents believe that the initiative was totally useless in terms of developing the indicator Comparison and Exchange

Therefore, it can be clearly seen that the majority of participants believes that the Educational Robotics project was useful for the development of the indicator "Comparison and Exchange"; more than 90% of respondents in fact attributes usefulness to the initiative in this sense, only one respondent considered the initiative as not useful.

Meeting the Timeframe

With respect to the question "How useful do you think the *e.DO Experience* initiative has been for the development of the indicator of Meeting the Timeframe?", it is possible to summarize the distribution of responses as follows:

- 31.5% of respondents felt that the initiative was very helpful in terms of developing the indicator Meeting the Timeframe
- 38.7% of respondents felt that the initiative was helpful in terms of developing the indicator Meeting the Timeframe
- 20.2% of respondents felt that the initiative was quite useful in terms of developing the indicator Meeting the Timeframe
- 8.1% of respondents felt that the initiative was indifferent in terms of indicator development Meeting the Timeframe

Figure 18. Meeting the Timeframe

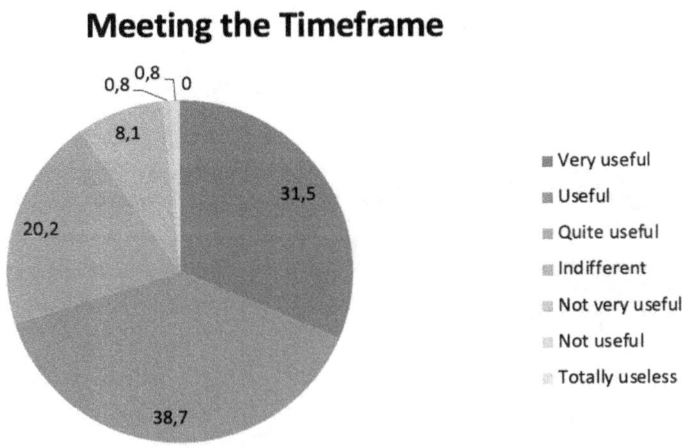

- 0.8% of respondents felt that the initiative was of little use in terms of developing the indicator Meeting the Timeframe
- 0.8% of respondents felt that the initiative was not helpful in terms of developing the indicator Meeting the Timeframe
- 0% of respondents felt that the initiative was totally unnecessary in terms of developing the indicator Meeting the Timeframe

Therefore, it can be clearly seen that the majority of participants believes that the Educational Robotics project was useful for the development of the " Meeting the Timeframe" indicator. In fact, 90% of respondents attribute usefulness in this sense, while only 2 participants do not consider the initiative as useful.

Celebrating Success

With respect to the question "How useful do you think the *e.DO Experience* initiative has been for the development of the indicator Celebrating Success?", it is possible to summarize the distribution of responses as follows:

- 27.4% of respondents felt that the initiative was very helpful in terms of developing the indicator Celebrating Success
- 33.9% of respondents felt that the initiative was helpful in terms of developing the indicator Celebrating Success
- 22.6% of respondents felt that the initiative was quite useful in terms of developing the indicator Celebrating Success

Figure 19. Distribution of responses related to the "Celebrating Success" item

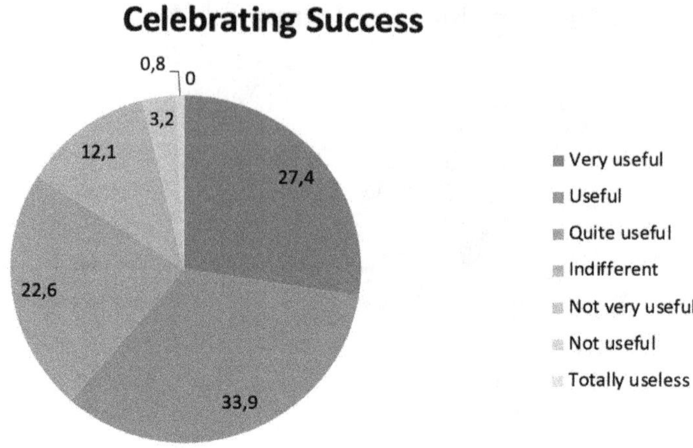

- 12.1% of respondents felt that the initiative was indifferent in terms of developing the indicator Celebrating Success
- 3.2% of respondents felt the initiative was of little use in terms of developing the indicator Celebrating Success
- 0.8% of respondents felt that the initiative was not helpful in terms of developing the indicator Celebrating Success
- 0% of respondents felt that the initiative was totally unnecessary in terms of developing the indicator Celebrating Success

Thus, it can be clearly seen that most participants felt that the Educational Robotics project was helpful in developing the "Celebrating Success" indicator.

The survey clearly highlighted the perspective of respondents, underscoring a widespread perception of usefulness with respect to the indicators that make up the "Value for New Citizenship" area, with the exception of the indicators "Equity and Justice", "Being Exemplary and Consistent", "Being Transparent and Sincere" and "Keeping Promises", for which usefulness is poorly perceived.

In this regard, two main considerations can be made. The scale used to measure the "Value for New Citizenship" dimension obtained, among all those considered in this Survey, the lowest score in terms of statistical reliability, reporting a Chronbach's α between .6 and .7, therefore barely acceptable. This implies that the measurement scale produced following the bibliographic review may not perfectly overlap the construct being measured and, consequently, not all items necessarily referred to the phenomenon. A further element to consider is that even in the interviews these indicators were made explicit only on rare occasions. Therefore, while the *e.DO Experience* is certainly an excellent means of developing a "Value for New Citizenship", it is equally **more useful for some areas (related to collaboration and challenge) than for other ones (related to elements of honesty and fairness).**

Empowerment

Initially, authors asked the respondent via a single item (5-step Likert Scale) whether participants had perceived value in terms of "Empowerment". The question was phrased as follows:

In your opinion, did the people who took part in the training initiative you helped to promote and manage perceive value in terms of "Empowerment" (Becoming aware of one's abilities, being resourceful and proactive)?

The question in object included 5 possible answers indicative of the presence or absence of a usefulness of the *e.DO Experience* in terms of Empowerment:

- Absolutely yes
- Probably yes
- Neither yes nor no
- Probably not
- Absolutely not

Specifically, the responding population thus rated the usefulness of *e.DO Experience* initiatives in terms of Empowerment:

- 41.1% believes that the *e.DO Experience* is absolutely useful for generating value in terms of Empowerment
- 56.6% believes that the *e.DO Experience* is likely to be useful in generating value in terms of Empowerment
- 2.4% believes that the *e.DO Experience* is neither useful nor unhelpful in generating value in terms of Empowerment
- 0% believes that the *e.DO Experience* is probably not useful for generating value in terms of Empowerment
- 0% believes that the *e.DO Experience* is completely unhelpful in generating value in terms of Empowerment

Table 4. Descriptive statistics related to the "Empowerment" area

Empowerment Value Perception	Frequency	Percentage
Absolutely yes	51	41.1
Probably yes	70	56.5
neither yes nor no	3	2.4
Probably not	0	0
Absolutely not	0	0

Table 4 and Figure 20 shows the descriptive statistics related to this area of value.

It can therefore be seen here, from a purely descriptive point of view just valid within our sample, that the majority of respondents believes that participants have created value in terms of Empowerment (41.1% absolutely yes, 56.5% probably yes). In this case, the answer does not appear to be influenced by the organization to which the respondent belongs to (t (122) = 1.192, p = .236). Next, authors sized the usefulness of the initiative in terms of "Empowerment" through a scale (5 items, alpha = .86; 7-point Likert). For each item that makes up the "Empowerment" scale, we report descriptive statistics:

Being Satisfied and Enthusiastic

With respect to the question "How useful do you think the *e.DO Experience* initiative was for the development of the indicator Being Satisfied and

Figure 20. Percentages of the distribution of value in the area "Empowerment"

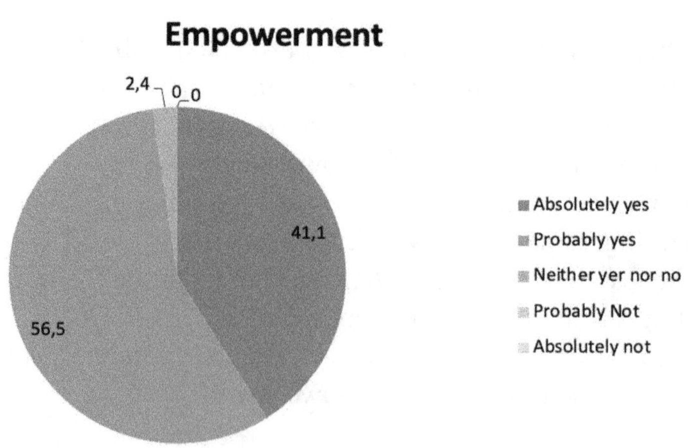

Enthusiastic?", it is possible to summarize the distribution of responses as follows:

- 53.2% of respondents felt that the initiative was very useful in terms of developing the indicator Being Satisfied and Enthusiastic
- 34.7% of respondents felt that the initiative was useful in terms of developing the indicator Being Satisfied and Enthusiastic
- 11.3% of respondents felt that the initiative was quite useful in terms of developing the indicator Being Satisfied and Enthusiastic
- 0.8% of respondents felt that the initiative was indifferent in terms of developing the indicator Being Satisfied and Enthusiastic
- 0% of respondents felt that the initiative was of little use in terms of the indicator Being Satisfied and Enthusiastic
- 0% of respondents felt that the initiative was not useful in terms of developing the indicator Being Satisfied and Enthusiastic
- 0% of respondents felt that the initiative was totally unnecessary in terms of developing the indicator Being Satisfied and Enthusiastic

Thus, it can be clearly seen that the majority of participants believes that the Educational Robotics project was useful for the development of the "Being Satisfied and Enthusiastic" indicator. In fact, no respondent indicated the initiative as little or no useful, only one indicated it as indifferent.

Figure 21. Distribution of responses related to the "Being Satisfied and Enthusiastic" item

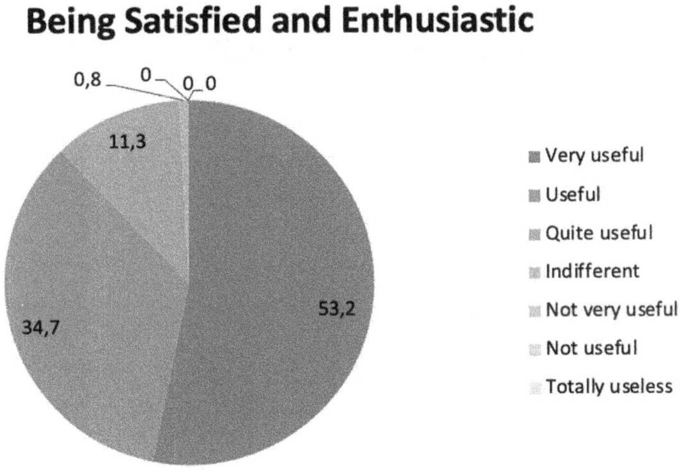

Being Aware of Your Own Qualities

With respect to the question "How useful do you think the *e.DO Experience* initiative has been for the development of the indicator Being Aware of Your Own Qualities?", it is possible to summarize the distribution of responses as follows:

- 32.2% of respondents felt that the initiative was very useful in terms of developing the indicator Being Aware of Your Own Qualities
- 53.2% of respondents felt that the initiative was useful in terms of developing the indicator Being Aware of Your Own Qualities
- 11.3% of respondents felt that the initiative was quite useful in terms of developing the indicator Being Aware of Your Own Qualities
- 3.2% of respondents felt that the initiative was indifferent in terms of developing the indicator Being Aware of Your Own Qualities
- 0% of respondents felt that the initiative was unhelpful in terms of the indicator Being Aware of Your Own Qualities
- 0% of respondents felt that the initiative was not useful in terms of developing the indicator Being Aware of Your Own Qualities
- 0% of respondents felt that the initiative was totally unnecessary in terms of developing the indicator Being Aware of Your Own Qualities

Therefore, it can be clearly seen that the majority of participants believes that the Educational Robotics project was useful for the development of

Figure 22. Distribution of responses related to the "beinng Aware of Your Own Qualities" item

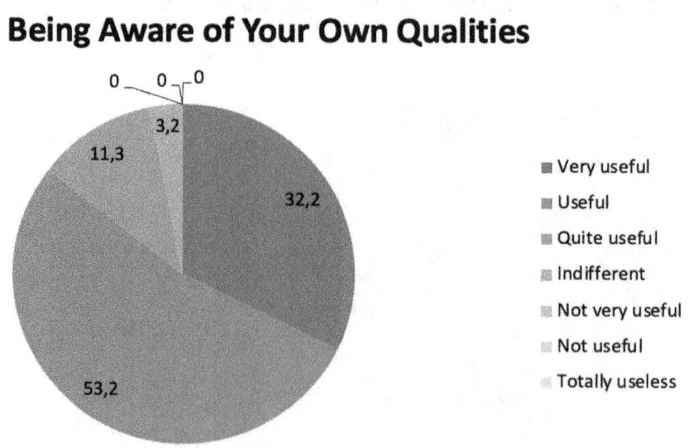

the indicator "Being Aware of Your Own Qualities". In fact, no respondent indicated the initiative as little or no useful, only four indicated it as indifferent.

Being Confident and Constructive

With respect to the question "How useful do you think the *e.DO Experience* initiative has been for the development of the indicator Being Confident and Constructive?", it is possible to summarize the distribution of responses as follows:

- 32.2% of respondents felt that the initiative was very useful in terms of developing the indicator Being Confident and Constructive
- 46.8% of respondents felt that the initiative was helpful in terms of developing the indicator Being Confident and Constructive
- 19.4% of respondents felt that the initiative was quite useful in terms of developing the indicator Being Confident and Constructive
- 1.6% of respondents felt that the initiative was indifferent in terms of developing the indicator Being Confident and Constructive
- 0% of respondents felt that the initiative was unhelpful in terms of the indicator Being Confident and Constructive
- 0% of respondents felt that the initiative was not helpful in terms of developing the indicator Being Confident and Constructive
- 0% of respondents felt that the initiative was totally unnecessary in terms of developing the indicator Being Confident and Constructive

Figure 23. Distribution of responses related to the "Being Confident annd Constructive" item

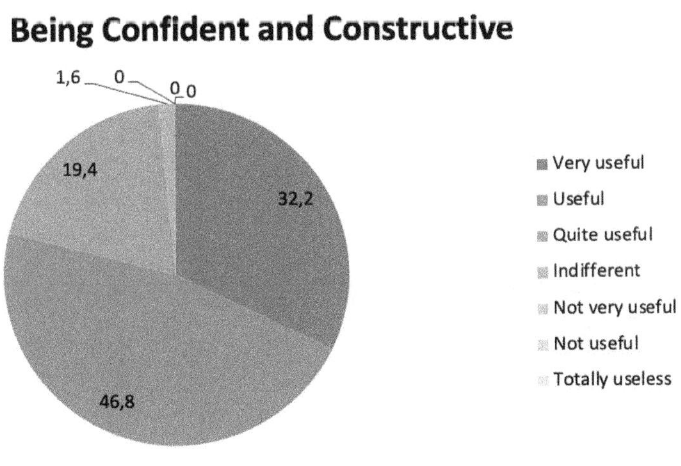

Thus, it can be clearly seen that the majority of participants believes that the Educational Robotics project was useful for the development of the "Being Confident and Constructive" indicator. In fact, no respondents indicated the initiative as little or no useful, only two indicated it as indifferent.

Being Passionate and Energetic

With respect to the question "How useful do you think the *e.DO Experience* initiative was for the development of the indicator Being Passionate and Energetic?", it is possible to summarize the distribution of responses as follows:

- 39.5% of respondents felt that the initiative was very helpful in terms of developing the indicator Being Passionate and Energetic
- 38.7% of respondents felt that the initiative was helpful in terms of developing the indicator Being Passionate and Energetic
- 18.5% of respondents felt that the initiative was quite helpful in terms of developing the indicator Being Passionate and Energetic
- 2.4% of respondents felt that the initiative was indifferent in terms of developing the indicator Being Passionate and Energetic
- 0.8% of respondents felt that the initiative was unhelpful in terms of the indicator Being Passionate and Energetic
- 0% of respondents felt that the initiative was not helpful in terms of developing the indicator Being Passionate and Energetic

Figure 24. Distribution of responses related to the "Being Passionate and Energetic" item

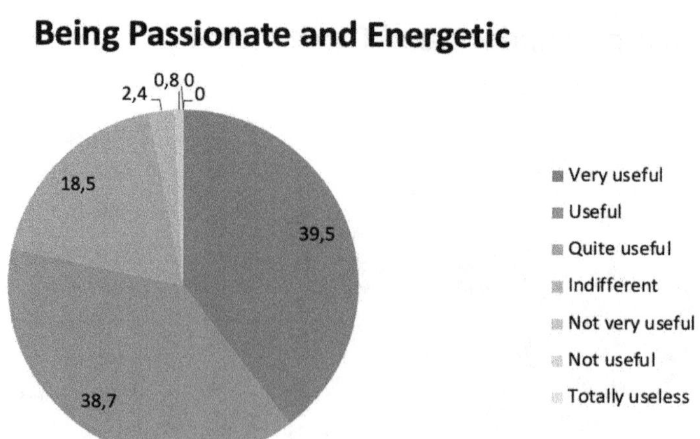

- 0% of respondents felt that the initiative was totally unnecessary in terms of developing the indicator Being Passionate and Energetic

Thus, it can be clearly seen that the majority of participants felt that the Educational Robotics project was beneficial to the development of the "Being Passionate and Energetic" indicator. Only three Participants reported non-positive responses.

Being Proactive and Resourceful

With respect to the question "How useful do you think the *e.DO Experience* initiative has been for the development of the indicator Being Proactive and Resourceful?", it is possible to summarize the distribution of responses as follows:

- 40.3% of respondents felt that the initiative was very useful in terms of developing the indicator Being Proactive and Resourceful
- 45.2% of respondents felt that the initiative was useful in terms of developing the indicator Being Proactive and Resourceful
- 10.5% of respondents felt that the initiative was quite useful in terms of developing the indicator Being Proactive and Resourceful
- 4% of respondents felt that the initiative was indifferent in terms of developing the indicator Being Proactive and Resourceful

Figure 25. Distribution of responses related to the "Being Proactive and Resourceful" item

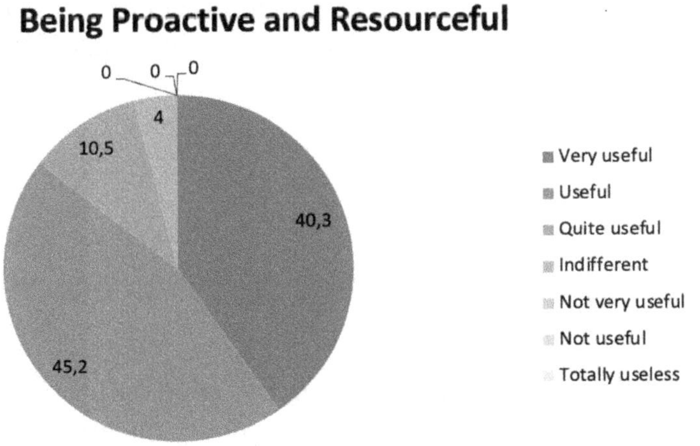

- 0% of respondents felt that the initiative was of little use in terms of the indicator Being Proactive and Resourceful
- 0% of respondents felt that the initiative was not useful in terms of developing the indicator Being Proactive and Resourceful
- 0% of respondents felt that the initiative was totally useless in terms of developing the indicator Being Proactive and Resourceful

Therefore, it can be clearly seen that most of the participants believe that the Educational Robotics project was useful for the development of the "Being Proactive and Resourceful" indicator. No respondent, in fact, considered the initiative to be of no use, only five indicated it as indifferent.

Overall, with respect to the Empowerment area, **all the different items that make up the measurement scale certify a widespread perception of usefulness for the *e.DO Experience*** concerning the development of this constituent element of Value for Participants.

Inspiration and Orienteering for Working

Authors first asked the respondent via a single item (5-Step Likert Scale) whether participants perceived value in terms of "Inspiration and Orienteering for Working". The question was phrased as follows:

In your opinion, did the people who took part in the training initiative you helped to promote and manage perceive value in terms of "INSPIRATION & ORIENTEERING FOR WORKING" (Generating Passion, Providing Professional Models through Engaging Experiences in Relationships with Technologies)?

The question in object included 5 possible answers indicative of the presence or absence of a usefulness of the *e.DO Experience* in terms of Inspiration & Orienteering for working:

- Absolutely yes
- Probably yes
- Neither yes nor no
- Probably not
- Absolutely not

Specifically, the responding population thus rated the usefulness of *e.DO Experience* initiatives in terms of Inspiration & Orienteering:

- 44.4% believes that the *e.DO Experience* is absolutely useful in generating value in terms of Inspiration & Orienteering for working
- 53.2% believes that the *e.DO Experience* is likely to be useful in generating value in terms of Inspiration & Orienteering for working
- 1.6% believes that the *e.DO Experience* is neither useful nor not useful for generating value in Inspiration & Orienteering for working
- 0.8% believes that the *e.DO Experience* is probably not useful for generating value in Inspiration & Orienteering for working
- 0% believes the *e.DO Experience* is absolutely not useful for generating value in Inspiration & Orienteering for working

Table 5 and Figure 26 show descriptive statistics related to this area of value.

It can therefore be seen here, from a purely descriptive point of view just valid within our sample, that the majority of respondents believes that participants created value in terms of Inspiration & Orienteering (44.4%

Table 5. Descriptive statistics related to the "Inspiration & Orienteering" area

Inspiration & Orienteering Value Perception	Frequency	Percentage
Absolutely yes	55	44.4
Probably yes	66	53.2
Neither yes nor no	2	1.6
Probably not	1	0.8
Absolutely not	0	0

Figure 26. Percentages of the distribution of value in the area "Inspiration & Orienteering"

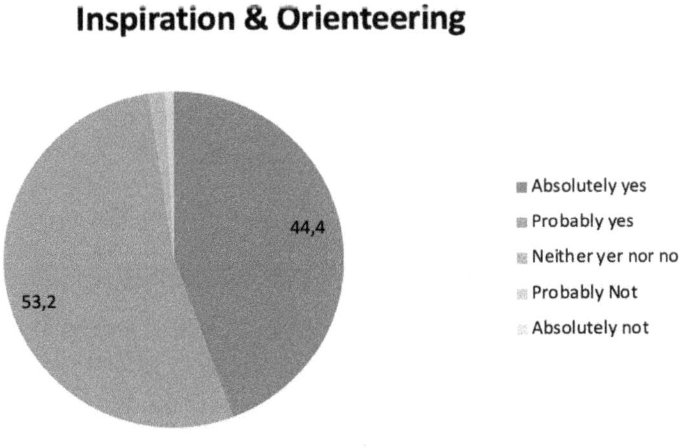

absolutely yes, 53.2% probably yes). In this case, the response does not appear to be influenced by the respondent's organization (t (122) = 1.145, p = .255).

Next, we sized the usefulness of the initiative in terms of "Inspiration & Orienteering" through a scale (5 items, alpha = .79; 7-point Likert). Of each item comprising the "Inspiration & Orienteering for working" scale, we report descriptive statistics:

Having Galvanizing Experiences

With respect to the question "How useful do you think the *e.DO Experience* initiative was for the development of the indicator Having Galvanizing Experiences?", it is possible to summarize the distribution of responses as follows:

- 50.8% of respondents felt that the initiative was very helpful in terms of developing the indicator Having Galvanizing Experiences
- 35.5% of respondents felt that the initiative was useful in terms of developing the indicator Having Galvanizing Experiences
- 12.1% of respondents felt that the initiative was quite helpful in terms of developing the indicator Having Galvanizing Experiences
- 0.8% of respondents felt that the initiative was indifferent in terms of developing the indicator Having Galvanizing Experiences
- 0.8% of respondents felt that the initiative was unhelpful in terms of the indicator Having Galvanizing Experiences

Figure 27. Distribution of responses related to the "Having Galvanizing Experiences" item

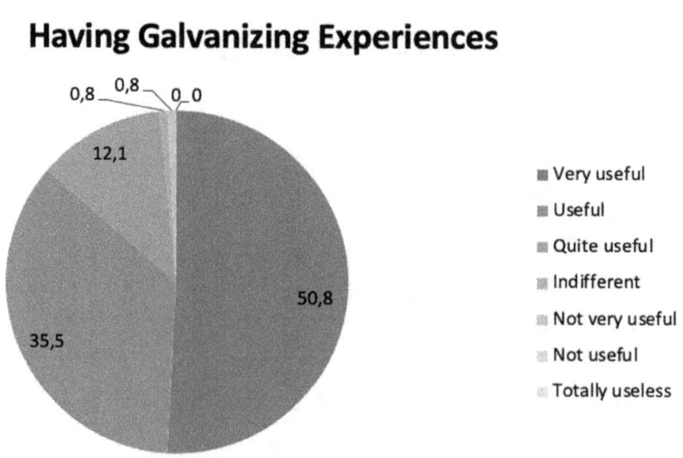

- 0% of respondents felt that the initiative was not helpful in terms of developing the indicator Having Galvanizing Experiences
- 0% of respondents felt that the initiative was totally unnecessary in terms of the indicator Having Galvanizing Experiences

Thus, it can be clearly observed that the majority of respondents felt that the Educational Robotics project was useful in developing the "Having Galvanizing Experiences" indicator. Only two respondents did not consider the initiative as useful in this regard.

Professional Model

With respect to the question "How useful do you think the *e.DO Experience* initiative has been in providing an inspirational Professional Model?", it is possible to summarize the distribution of responses as follows:

- 37.9% of respondents felt that the initiative was very much about offering an inspirational Professional Model
- 37.1% of respondents felt the initiative was helpful in providing an inspirational Professional Model
- 20.2% of respondents felt that the initiative was useful enough to offer an inspirational Professional Model
- 4% of respondents felt that the initiative was indifferent to offering an inspirational Professional Model

Figure 28. Distribution of responses related to the "Professional Model" item

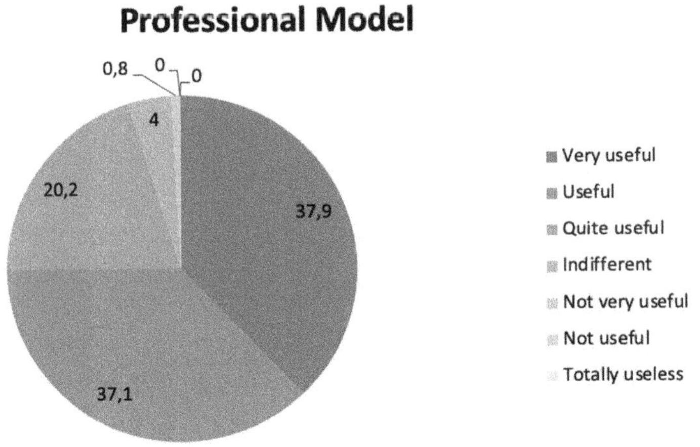

- 0.8% of respondents felt that the initiative did little to offer an inspirational Professional Model
- 0% of respondents felt that the initiative was not helpful in providing an inspirational Professional Model
- 0% of respondents felt that the initiative was totally unhelpful in providing an inspirational Professional Model

Thus, it can be clearly seen that the majority of respondents felt that the Educational Robotics project was useful in developing the "Professional Model" indicator. In fact, no respondents indicated the initiative as not useful, only one as not very useful, and five as indifferent.

Professional Trajectories

With respect to the question "How useful do you think the *e.DO Experience* initiative has been in offering Professional Trajectories?", it is possible to summarize the distribution of responses as follows:

- 37.1% of respondents felt that the initiative was very helpful in providing Professional Trajectories
- 43.5% of respondents felt the initiative was helpful in providing Professional Trajectories
- 12.9% of respondents felt that the initiative was quite helpful in offering Professional Trajectories

Figure 29. Distribution of responses related to the "Professional Trajectories" item

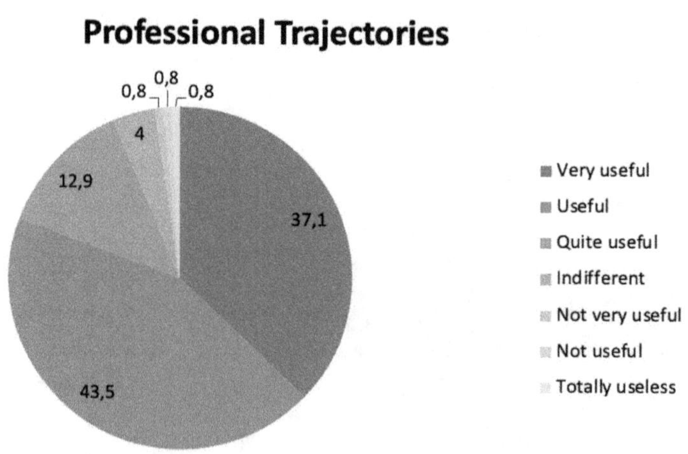

- 4% of respondents felt that the initiative was indifferently useful in offering Professional Trajectories
- 0.8% of respondents felt that the initiative was of little use in offering Professional Trajectories
- 0.8% of respondents felt that the initiative was not helpful in offering Professional Trajectories
- 0.8% of respondents felt that the initiative was totally unhelpful in offering Professional Trajectories

Thus, it can be clearly seen that most Participants felt that the Educational Robotics project was useful for the development of the Professional Trajectories indicator. More than 80% of participants felt the initiative was useful in this regard.

World of Work Awareness

With respect to the question "How useful do you think the *e.DO Experience* initiative has been in developing a greater Awareness of the World of Work?", it is possible to summarize the distribution of responses as follows:

- 30.6% of respondents felt that the initiative was very helpful in developing a greater World of Work Awareness
- 40.3% of respondents felt that the initiative was helpful in developing greater World of Work Awareness

Figure 30. Distribution of responses related to the "World of Work Awareness" item

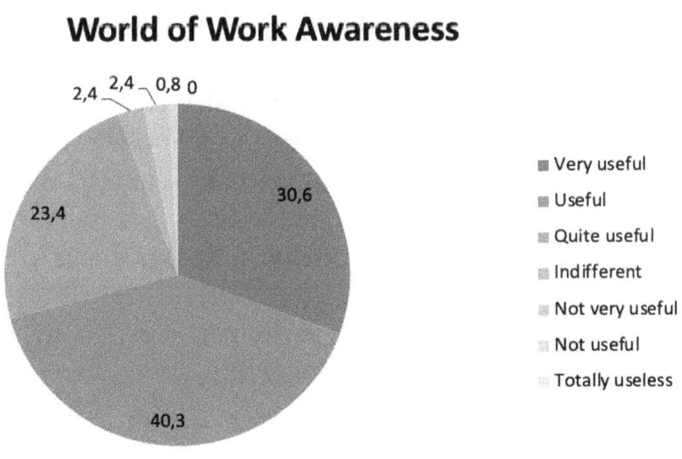

- 23.4% of respondents felt that the initiative was quite helpful in developing a greater World of Work Awareness
- 2.4% of respondents felt that the initiative was indifferently helpful in developing greater World of Work Awareness
- 2.4% of respondents felt that the initiative was of little use in developing World of Work Awareness
- 0.8% of respondents felt that the initiative was not helpful in developing World of Work Awareness
- 0% of respondents felt that the initiative was totally unhelpful in developing greater World of Work Awareness

Thus, it can be clearly seen that most participants felt that the Educational Robotics project was useful in developing greater "World of Work Awareness."

Awareness of One's Strengths and Weaknesses

With respect to the question "How useful do you think the *e.DO Experience* initiative has been in developing a greater Awareness of One's Strengths and Weaknesses?", it is possible to summarize the distribution of responses as follows:

- 25.8% of respondents felt that the initiative was very helpful in developing a greater Awareness of One's Strengths and Weaknesses
- 43.5% of respondents felt that the initiative was helpful in developing a greater Awareness of One's Strengths and Weaknesses
- 26.6% of respondents felt that the initiative was quite helpful in developing a greater Awareness of One's Strengths and Weaknesses
- 3.2% of respondents felt that the initiative was indifferently helpful in developing a greater Awareness of One's Strengths and Weaknesses
- 0.8% of respondents felt that the initiative was of little use in developing a greater Awareness of One's Strengths and Weaknesses
- 0% of respondents felt that the initiative was not helpful in developing a greater Awareness of One's Strengths and Weaknesses
- 0% of respondents felt that the initiative was totally unhelpful in developing a greater Awareness of One's Strengths and Weaknesses

Thus, it can be clearly seen that most participants felt that the Educational Robotics project was useful in developing greater "Awareness of One's Strengths and Weaknesses."

Figure 31. Distribution of responses related to the "Awareness of One's Strenghts and Weaknesses" item

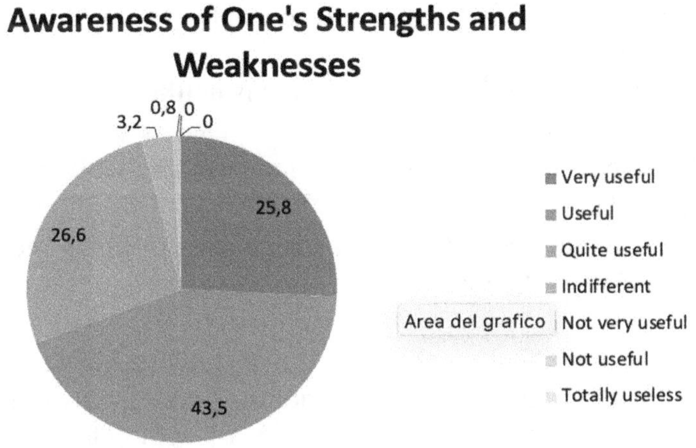

Value for Company

This level of the value generated by Educational Robotics initiatives was explored through the use of a single item (5-point Likert scale) and a scale referring to the main indicators of the "Value for Company" item (alpha = 0.89; Item number: 7; 7-point Likert scale). Initially, authors asked the respondent through a single item (5-step Likert Scale) whether the participants had perceived value in terms of "Value for Company". The question was phrased as follows:

In your opinion, did the Comau Academy educational robotics initiative that you helped to promote and manage have any specific value to your Company?

The question in object included 5 possible answers indicative of the presence or absence of a usefulness of the *e.DO Experience* in terms of Value for Company:

- Absolutely yes
- Probably yes
- Neither yes nor no
- Probably not
- Absolutely not

Specifically, the responding population thus rated the usefulness of *e.DO Experience* project in terms of Value for Company:

- 52.4% believes that the *e.DO Experience* is absolutely useful in generating Value for Company
- 37.9% believes the *e.DO Experience* is likely to be useful in generating Value for Company
- 8.9% believes that the *e.DO Experience* is neither useful nor unhelpful in generating Value for Company
- 0.8% believes that the *e.DO Experience* is probably not useful for generating Value for Company
- 0% believes that the *e.DO Experience* is absolutely not useful to generate Value for Company

Table 6 and Figure 32 show the descriptive statistics related to this item. Therefore, from a purely descriptive point of view just valid within our sample, it can be seen here that the majority of respondents believes that participants have perceived a Value for Company resulting from Educational Robotics project (52.4% absolutely yes, 37.9% probably yes). In this case, the

Table 6. Frequency and percentage of Company Value Perception in terms of usefulness

Company Value Perception	Frequency	Percentage
Absolutely yes	65	52.4
Probably yes	47	37.9
neither yes nor no	11	8.9
Probably not	1	0.8
Absolutely not	0	0

Figure 32. Descriptive statistics of usefulness perceived in terms of Value for Company

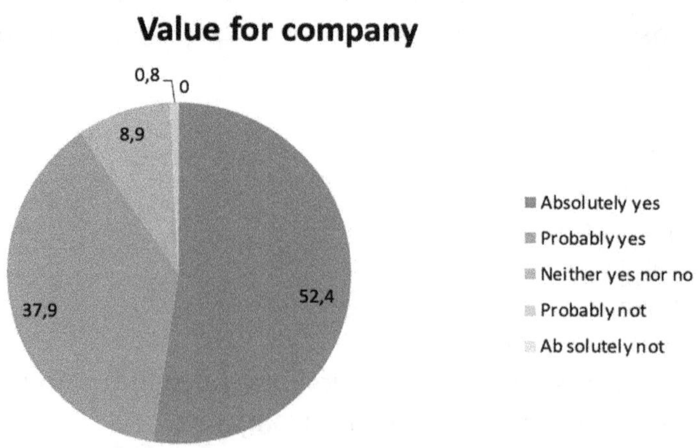

response does not appear to be influenced by the respondent's organization (t (122) = .305, p = .761).

With the second item is measured, in percentage terms, the weight that the various respondents attribute to the seven sub-dimensions of the "Value for Company" area. Next, the item is reported as the respondents read it:

Consider 100 as the total value of the Comau Academy educational robotics initiative that you helped promote and manage. As a percentage, how much do you think the following components impact value sharing with the Company? (indicate number of responses, total must equal 100):

- "Core Business"
- "Brand"
- "Employee Branding"
- "Formal CSR"
- "CSR True Commitment"
- "Open to Innovation"
- "Employee Exposure".

The descriptive analyses allowed to define the average score, in percentages, for each of the following areas. Table 7 is provided, in which the mean of each item is reported.

From a purely descriptive point of view, the area indicated as most relevant in terms of creating Value for Company is "Brand" (20,13%), followed by "Core Business" (17,02%), "Open to innovation" (15,53%), "Employee Branding" (14,01%), "CSR true commitment" (13,74%), "Employee exposure" (11,30%) and finally the item "Formal CSR" (8,27%). Descriptive percentages of the distribution are briefly shown in the chart below:

These first descriptive indications highlight a dual nature of the initiative: if on the one hand it generates value in commercial terms in relation to the

Table 7. Average score for each area of the level "Value for Company"

Areas	Media
Core Business	17,02%
Brand	20,13%
Employee Branding	14,01%
Formal CSR	8,27%
CSR True Commitment	13,74%
Open to innovation	15,53%
Employee Exposure	11,30%

Figure 33. Descriptive percentages of the distribution of value in the areas

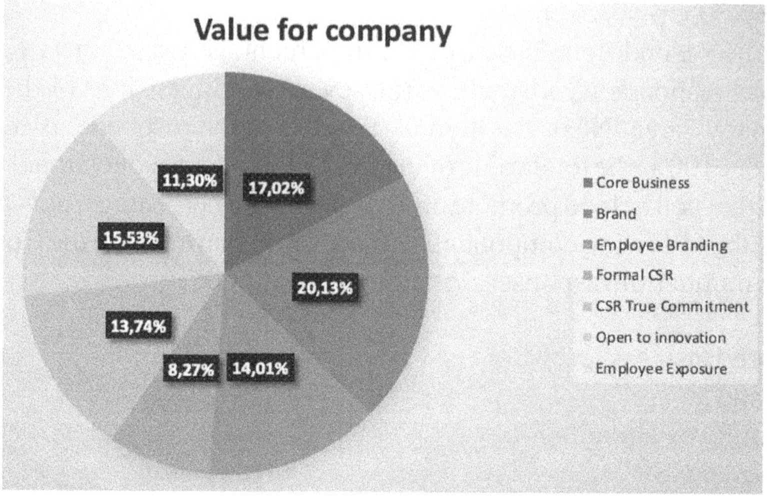

development of the brand and the core business, on the other hand it activates processes of organizational growth supported by a responsible interest towards the external community and internal employees.

Next, authors sized the usefulness of the project in terms of "Value to the Company" through a scale (5 items, alpha = .67; 7-point likert). The different items then made it possible to size the utility perceived by respondents in terms of "Core Business", "Brand", "Employee Branding", "Formal CSR", "CSR True commitment", "Open to Innovation" and "Employee Exposure".

Core Business

With respect to the question "How useful do you think the *e.DO Experience* initiative has been for the development of the Core Business?", it is possible to summarize the distribution of responses as it follows:

- 29% of respondents felt that the initiative was very useful for Core Business development
- 37.9% of respondents felt that the initiative was useful for Core Business development
- 22.6% of respondents felt the initiative was quite useful for Core Business development
- 8.1% of respondents felt that the initiative was indifferently useful for Core Business development

Figure 34. Distribution of responses related to the "Core Business" item

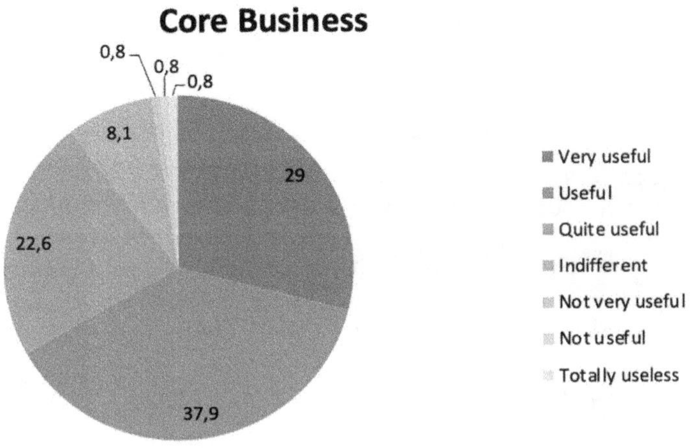

- 0.8% of respondents felt that the initiative was of little use for Core Business development
- 0.8% of respondents felt that the initiative was not useful for Core Business development
- 0.8% of respondents felt that the initiative was totally unnecessary for Core Business development

It can be clearly seen, therefore, that the majority of the participants believes that the Educational Robotics project was perceived as useful for the development of the "Core Business" indicator. In fact, 90% of participants felt the initiative was useful in this regard.

Brand

With respect to the question "How useful do you think the *e.DO Experience* initiative has been for the development of the Brand?", it is possible to summarize the distribution of responses as follows:

- 37.1% of respondents felt that the initiative was very useful for Brand development
- 37.9% of respondents felt the initiative was helpful for Brand development
- 16.9% of respondents felt the initiative was quite useful for Brand development

Figure 35. Distribution of responses related to the "Brand" item

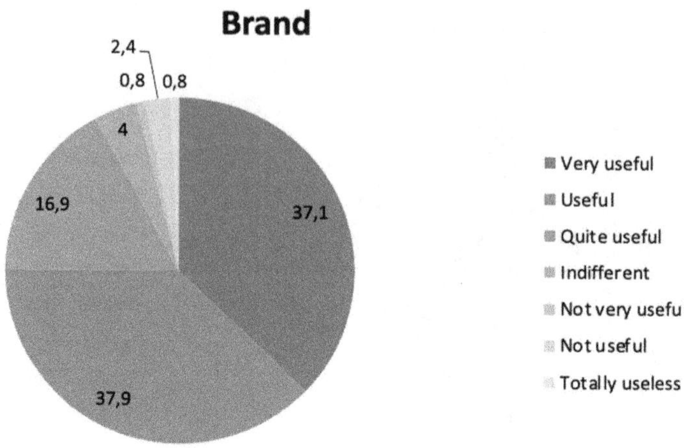

- 4% of respondents felt that the initiative was indifferently useful for Brand development
- 0.8% of respondents felt that the initiative was of little use in Brand development
- 2.4% of respondents felt that the initiative was not helpful for Brand development
- 0.8% of respondents felt that the initiative was totally unnecessary for Brand development

Therefore, it can be clearly seen that most of the Participants believe that the Educational Robotics project was perceived as useful for the development of the "Brand" indicator, perceived as usual by more than 90% of the respondents.

Employee Branding

With respect to the question "How useful do you think the *e.DO Experience* initiative has been for the development of the Employee Branding indicator?", it is possible to summarize the distribution of responses as follows:

- 29% of respondents felt that the initiative was very helpful in developing the Employee Branding indicator
- 32.3% of respondents felt that the initiative was helpful in developing the Employee Branding indicator

Figure 36. Distribution of responses related to the "Employee Branding" item

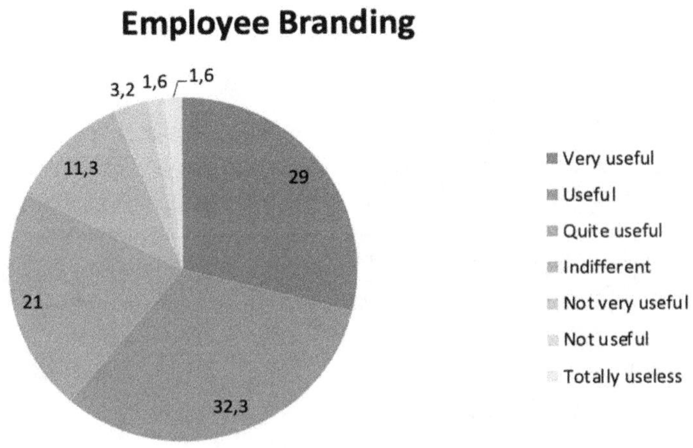

- 21% of respondents felt that the initiative was quite useful in developing the Employee Branding indicator
- 11.3% of respondents felt that the initiative was indifferently useful for the development of the Employee Branding indicator
- 3.2% of respondents felt that the initiative was of little use in developing the Employee Branding indicator
- 1.6% of respondents felt that the initiative was not helpful in developing the Employee Branding indicator
- 1.6% of respondents felt that the initiative was totally unnecessary for the development of the Employee Branding indicator

Thus, it can be clearly seen that the majority of participants (about 85%) felt that the Educational Robotics project was useful in developing the Employee Branding indicator.

Formal CSR

With respect to the question "How useful do you think the *e.DO Experience* initiative has been for the development of Formal CSR?", it is possible to summarize the distribution of responses as follows:

- 20.2% of respondents felt that the initiative was very helpful in developing Formal CSR

Figure 37. Distribution of responses related to the "Formal CSR" item

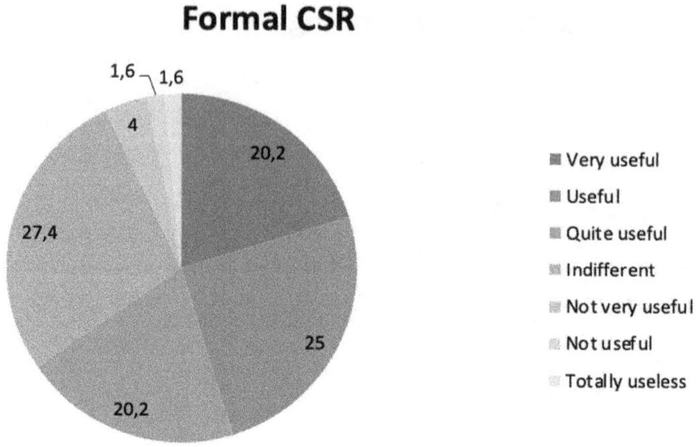

- Twenty-five percent of respondents felt the initiative was helpful in developing Formal CSR
- 20.2% of respondents felt that the initiative was quite helpful in developing Formal CSR
- 27.4% of respondents felt that the initiative was indifferently helpful in developing Formal CSR
- 4% of respondents felt that the initiative was of little use in developing Formal CSRs
- 1.6% of respondents felt that the initiative was not helpful in developing Formal CSRs
- 1.6% of respondents felt that the initiative was totally unnecessary for the development of Formal CSR

Thus, it can be clearly seen that just over 60% of the participants felt that the Educational Robotics project was useful in developing the "Formal CSR" indicator. Of all the indicators related to Value for Company, this is the one that received the lowest score. This indication confirms the hypothesis already developed in the interviews.

CSR True Commitment

With respect to the question "How useful do you think the *e.DO Experience* initiative has been for the development of CSR True Commitment?", it is possible to summarize the distribution of responses as follows:

Figure 38. Distribution of responses related to the "CSR True Commitment" item

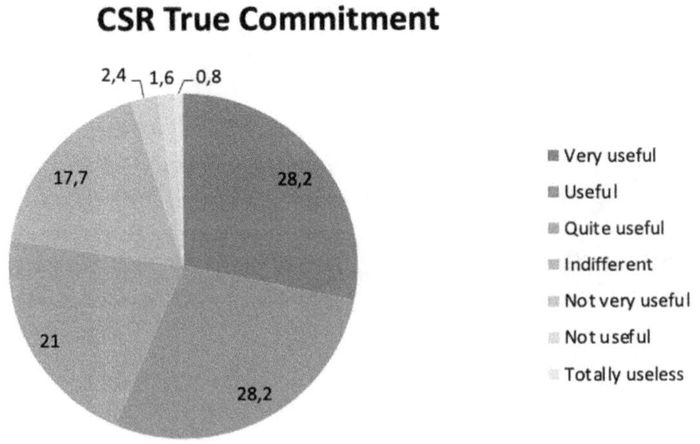

- 28.2% of respondents felt that the initiative was very helpful in developing CSR True Commitment
- 28.2% of respondents felt the initiative was helpful in developing CSR True Commitment
- 21% of respondents felt the initiative was quite helpful in developing CSR True Commitment
- 17.7% of respondents felt that the initiative was indifferently helpful in developing CSR True Commitment
- 2.4% of respondents felt that the initiative was of little use in developing CSR True Commitment
- 1.6% of respondents felt that the initiative was not helpful in developing CSR True Commitment
- 0.8% of respondents felt that the initiative was totally unhelpful in developing CSR True Commitment

Thus, it can be clearly seen that less than 80% of participants felt that the Educational Robotics project was useful in developing the "CSR True Commitment" indicator.

Open to Innovation

With respect to the question "How useful do you think the *e.DO Experience* initiative has been for the development of the Open to Innovation indicator?", it is possible to summarize the distribution of responses as follows:

Figure 39. Distribution of responses related to the "Open to Innovation" item

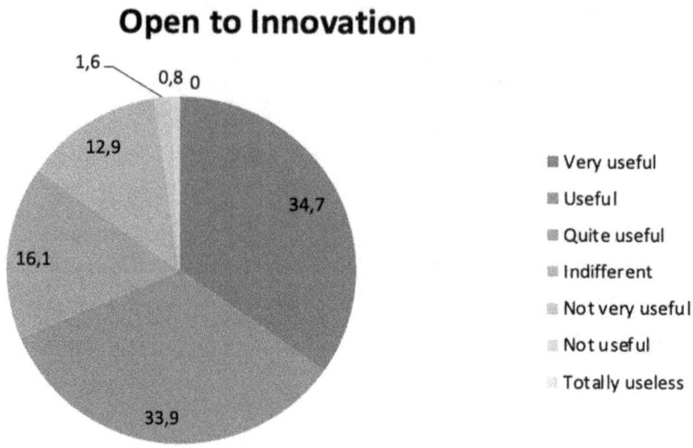

- 34.7% of respondents felt that the initiative was very helpful in developing the Open to Innovation indicator
- 33.9% of respondents felt that the initiative was helpful in developing the Open to Innovation indicator
- 16.1% of respondents felt that the initiative was quite helpful in developing the Open to Innovation indicator
- 12.9% of respondents felt that the initiative was indifferently useful for the development of the Open to Innovation indicator
- 1.6% of respondents felt that the initiative was of little use in developing the Open to Innovation indicator
- 0.8% of respondents felt that the initiative was not helpful in developing the Open to Innovation indicator
- 0% of respondents felt that the initiative was totally unnecessary for the development of the Open to Innovation indicator

Thus, it can be clearly seen that the majority of participants felt that the Educational Robotics project was beneficial to the development of the "Open to Innovation" indicator.

Employee Exposure

With respect to the question "How useful do you think the *e.DO Experience* initiative has been for the development of the Employee Exposure indicator?", the distribution of responses can be summarized as follows:

Figure 40. Distribution of responses related to the "Employee Exposure" item

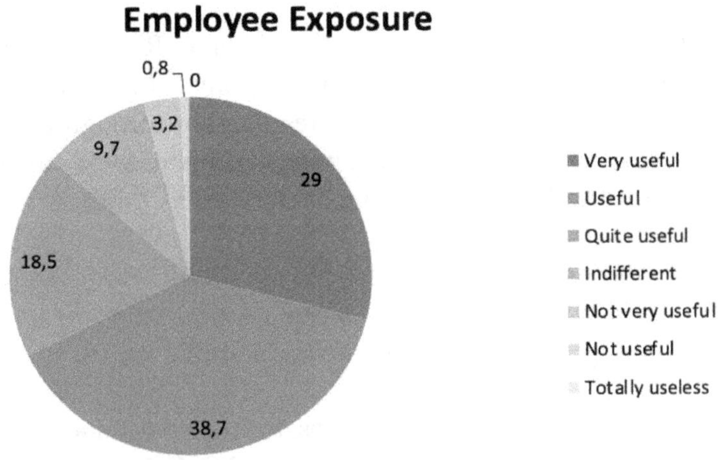

- 29% of respondents felt that the initiative was very helpful in the development of the Employee Exposure indicator
- 38.7% of respondents felt that the initiative was helpful in the development of the Employee Exposure indicator
- 18.5% of respondents felt that the initiative was quite useful for the development of the Employee Exposure indicator
- 9.7% of respondents felt that the initiative was indifferently useful for the development of the Employee Exposure indicator
- 3.2% of respondents felt that the initiative was of little use in the development of the Employee Exposure indicator
- 0.8% of respondents felt that the initiative was not helpful in the development of the Employee Exposure indicator
- 0% of respondents felt that the initiative was totally unnecessary for the development of the Employee Exposure indicator

Thus, it can be clearly observed that most participants felt that the Educational Robotics project was useful in developing the "Employee Exposure" indicator.

The Survey has clearly highlighted the perspective of respondents, underlining a **widespread perception of usefulness with respect to the indicators that make up the "Value for Company" area**, with the exception of the item relating to Formal CSR, for which usefulness is little perceived or lower than normally assumed.

Value for Territory

This level of the value generated by Educational Robotics project was explored through the use of two single items (5-point Likert scale; percentage distribution). At first, authors asked the respondent through a single item (5-step Likert Scale) whether the participants had perceived value in terms of "Value for Territory". The question was phrased as follows:

In your opinion, did the Comau Academy educational robotics initiative that you helped to promote and manage have any specific value for the area that hosted it and, more generally, for the outside world?

The question in object included 5 possible answers indicative of the presence or absence of a usefulness of the *e.DO Experience* in terms of Value for Territory:

- Absolutely yes
- Probably yes
- Neither yes nor no
- Probably not
- Absolutely not

Specifically, the responding population rated the usefulness of *e.DO Experience* initiatives in terms of their value to the area as follows:

- 44.4% believes that the *e.DO Experience* is absolutely useful for generating value for the area
- 53.2% believes that the *e.DO Experience* is likely to be useful in generating value for the area
- 1.6% believes that the *e.DO Experience* is neither useful nor unhelpful in generating value for the area
- 0.8% believes that the *e.DO Experience* is probably not useful for generating value for the area
- 0% believes that the *e.DO Experience* is absolutely not useful for generating Value for Territory

Table 8 and Figure 41 show descriptive statistics related to this item.

Table 8. Frequency and percentage of Territory Value Perception in terms of usefulness

Territory Value Perception	Frequency	Percentage
Absolutely yes	55	44.4
Probably yes	66	53.2
neither yes nor no	2	1.6
Probably not	1	0.8
Absolutely not	0	0

Figure 41. Descriptive statistics of usefulness perceived in terms of Value for Territory

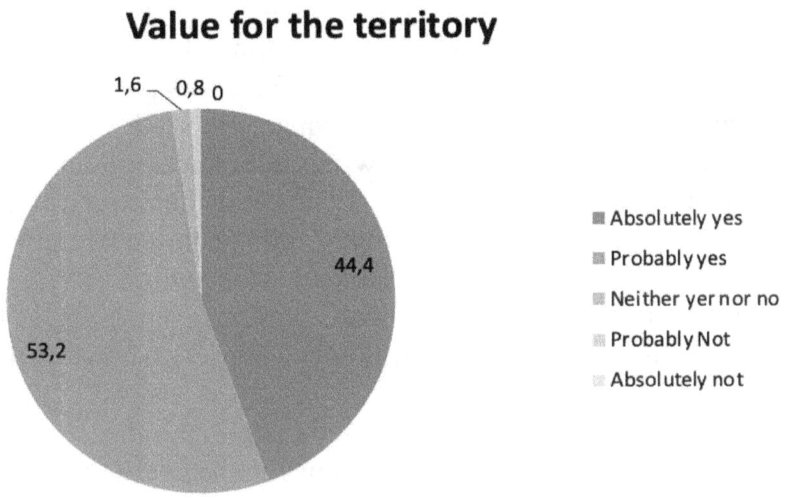

With the second item, we measured the weight that the various respondents attribute to the four areas of the "Value for Territory" level in percentage terms. Below, the item is reported as the respondents read it:

Consider 100 as the total value of the Comau Academy educational robotics initiative that you helped to promote and manage. As a percentage, how much do you think the following components impact value sharing with the local area? (indicate number of responses, total must equal 100):

- Creation and dissemination of Culture
- Educational Innovation
- Contribution to the Creation of skills required by the World of Work
- Contribution to the Creation of New Ecosystems

Table 9. Average score for each area of the level "Value for Territory"

Areas	Media	Std. deviation
Creation and dissemination of culture	25.85%	12.51
Educational Innovation	33.02%	12.89
Contribution to the creation of skills required by the world of work	26.77%	11.78
Contribution to the creation of new ecosystems	14.27%	7.83

The descriptive analysis allowed us to define the average score, in percentages, for each of the following areas: Table 9 is provided, in which the mean and standard deviation of each item are reported.

From a purely descriptive point of view, the area indicated as most relevant in terms of creating Value for Territory is "Educational Innovation" (33.02%), followed by "Contribution to the creation of skills required by the World of Work" (26.77%), "Creation and Dissemination of culture" (25.85%) and finally the item "Contribution to the Creation of New Ecosystems" (14.27%).

Descriptive percentages of the distribution are briefly shown in the chart below.

While these descriptive statistics clearly highlight the **considerable weight that "Educational Innovation" plays** in the creation of Value for Territory, it can be noted that respondents also consider central the aspects

Figure 42. Descriptive percentages of the distribution of value in the areas

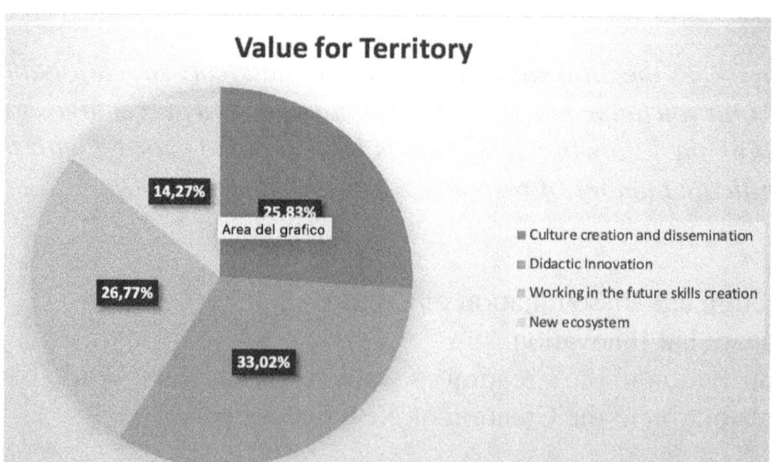

of Dissemination of Culture, Creation of skills enabling work in the future and New Ecosystems (Pinto et al., 2021).

Using the descriptive statistics above, it is possible to size the four areas according to a valid average percentage for the sample used. Authors then wanted to verify whether being a Comau Manager or a Partner had statistically significant differences with respect to the areas mentioned above; inferential statistics show that for the items "Culture Creation and dissemination" (t(122) = 1.838, p = .069), "Educational Innovation" (t (122) = .997, p = .340), " Creation of skills required by the World of Work" (t (122) = 1.35, p = .180) and "Creation of New Ecosystems" (t (122) = .539, p = .591) do not show statistically significant differences related to the organization they belong to (1 = Comau; 2 = Partner Company). Consequently, it can be inferred that **the various items referring to the fraction of Value for Territory have the same weight for both Comau Managers and Partners**.

Correlational Analysis

In concluding the quantitative study, authors wanted to test the existence and strength of **the link between the indicator for Value for Company and the different scales used to size the Value for Participants**. Table 10 summarizes the results of the correlational study.

What emerges from this inferential study is the presence of **statistically significant positive correlations among all the different scales considered**.

Table 10. Correlational study

		Value for Company	Skills Creation	New Citizenship	Empowerment	Inspiring Orienteering
Value for Company	Pearson Correlation	1	,398**	,417**	,395**	,550**
	Sig		,000	,000	,000	,000
	N	124	124	124	124	124
Skills Creation	Pearson Correlation	,398**	1	,435**	,280**	,469**
	Sig	,000		,000	,002	,000
	N	124	124	124	124	124
New Citizenship	Pearson Correlation	,417**	,435**	1	,490**	,550**
	Sig	,000	,000		,000	,000
	N	124	124	124	124	124
Empowerment	Pearson Correlation	,395**	,280**	,490**	1	,434**
	Sig	,000	,002	,000		,000
	N	124	124	124	124	124
Inspiring Orienteering	Pearson Correlation	,550**	,469**	,550**	,434**	1
	Sig	,000	,000	,000	,000	
	N	124	124	124	124	124

The perception of Value for Company correlates positively with value in terms of Skill Creation (r124, = .40; $p \leq .001$), Value for New Citizenship (r124, = .42; $p < .001$), Empowerment (r124, = .40; $p < .001$) and Inspiration and Orienteering for Working (r124, = .55; $p < .001$). The different scales relating to Value for Participants are all correlated: Skill Creation correlates positively with Value for New Citizenship (r124, = .44; $p \leq .001$), Empowerment ($r(124) = .28$, $p = .002$) and Inspiration and Orienteering for Working (r124, = .47; $p < .001$); Value for New Citizenship correlates positively with Empowerment (r124, = .49; $p \leq .001$) and Inspiration and Orienteering for Working (r124, = .55; $p < .001$), while Empowerment and Inspiration and Orienteering for Working correlate with each other (r124, = .43; $p < .001$).

CONCLUSION

The Survey offers a series of confirmations with respect to the theoretical hypotheses developed in the previous phases: the *e.DO Experience* is perceived by managers and partners as an initiative capable of **generating Value for Participants, Company and Territory**.

The **Skill Creation** area is considered **the most relevant for the development of Value for Participants**, while the area of **Educational Innovation** is considered **the most significant for the Value for Territory**. On the other hand, correlational analysis shows that **the various scales relating to Value for Participants are positively correlated with the Value for Company**: this means that as the perception of Value for Participants increases, so does the perception of Value for Company.

In this scenario, the initial assumption that the *e.DO Experience* is configured as a way of sharing value (Porter & Kramer, 2011; 2019) between the Company, Participants and the Territory, seems to be confirmed.

REFERENCES

Fregnan, E., Ivaldi, S., & Scaratti, G. (2020). HRM 4.0 and New Managerial Competences Profile: The COMAU Case. *Frontiers in Psychology*, 11.

Pinto, D., Scaratti, G., & Fregnan, E. (2021). *Learning Agility 4.0: ecosistemi e trasformazione culturale*. Pearson.

Porter, M., & Kramer, M. R. (2011). Creare valore condiviso. *Harvard Business Review Italia*, *1*(2), 68–84.

Porter, M. E., & Kramer, M. R. (2019). Creating shared value. In *Managing sustainable business* (pp. 323–346). Springer. doi:10.1007/978-94-024-1144-7_16

Richards, L., & Morse, J. M. (2013). *Readme first for a user's guide to qualitative methods*. Sage.

Chapter 8
Learning in the Technological Contexts:
Evaluation of the Final Recipients

ABSTRACT

In this last chapter, the authors explore the perspective of those who participated in the Comau e.DO Experience Project, with the aim of verifying what emerged previously and using semiotic analysis. Twenty primary school children, 20 from first grade secondary school, and 20 second grade secondary school students were randomly selected from the workshop participants, foreshadowing the activation of three focus groups for each school level. For primary school children, the central element was the relationship of familiarity with modern technological tools; for secondary school students (10-12 years old), the pivotal aspect is the balance to be achieved between the human being and the machine. Finally, for secondary school students, the pivotal dimension concerns the technology used and the reference to the processes of innovation and transformation connected to it.

INTRODUCTION

The innovative experience of the *e.DO Experience* Project is part of the new educational trajectories oriented towards the use of robotics. As we have seen in the previous chapters, **the use of robots as artifacts and mediators of learning in educational contexts of various levels has triggered new and**

DOI: 10.4018/978-1-7998-8653-2.ch008

promising scenarios in the relationship between people and technologies, opening new perspectives for future generations.

In particular, the *e.DO Experience* has involved groups of students attending primary school, lower secondary and upper secondary secondary schools, all engaged in an immersive and intense laboratory path, in which the fears of inability and elements of skepticism are mixed with interest, pleasure and commitment to the educational proposal experienced. The *Robotic Learning Lab initiative* envisaged (see Chapter 4) a training situation called *e.DO Experience,* in which the participants faced, after a theoretical introduction, a series of activities divided into groups, with the task of solving one or more challenges, for instance problems contextualized in real situations. To face these tests, the students, accompanied by a Comau tutor, had the following tools at their disposal: their own school knowledge, the notions learned during the workshop, the possibility to collaborate with others and the e.DO robot (the robot specifically produced by Comau to carry out educational activities). The tests and the topics addressed were calibrated according to the ages of the participants and covered not only content related to STEM disciplines (Science, Technology, Engineering, Mathematics), but also aspects related to the humanities (economics, psychology, and sociology), addressed through a setting that stimulated the creativity and active participation of the students. In partnership with MIUR (Ministry for University and Research), about 3,000 students were involved in the *Robotic Learning Lab*, equally distributed between primary school and secondary school, together with their teachers. The objectives of the workshop concerned the acquisition of a basic knowledge of the world of robotics, the learning of some elements of the disciplines involved and the sensitization towards skills characterizing the new contemporary scenario, illustrated in the previous chapters (Deal with Human; Deal with Technologies; Agility; Engagement; Collaboration; Interdisciplinarity; Innovation).

In order to grasp the meanings attributed to the *e.DO Experience,* the use of **focus groups** was adopted, functional to a **semiotic, narrative and conversational reconnaissance** of the training situation experienced, soliciting a process of social construction of the value attributed to what happened in the *Robotic Learning Lab.* In particular, **the representations of the participants were investigated with respect to the solicited capabilities and with reference to the success factors** that made the training experience effective and interesting.

20 primary school children, 20 from lower secondary school and 20 upper secondary school students were randomly selected from the workshop participants, foreshadowing the activation of three focus groups for each school

level. One researcher prompted the discussion by proposing stimuli to the participants related to a defined area of interest, while a second researcher monitored the process through an audio recording and the writing of field notes. The stimuli proposed within the focus groups were the following ones:

1. Firstly, they were asked to **graphically represent their training experience** and then to **describe what they had drawn**, subsequently to **identify two adjectives that effectively described the salient features**;
2. Secondly, **a semantic differential** was proposed to facilitate the **positioning** of participants with respect to certain dimensions, related to the requested capabilities.

Figure 1. Focus group, part 1

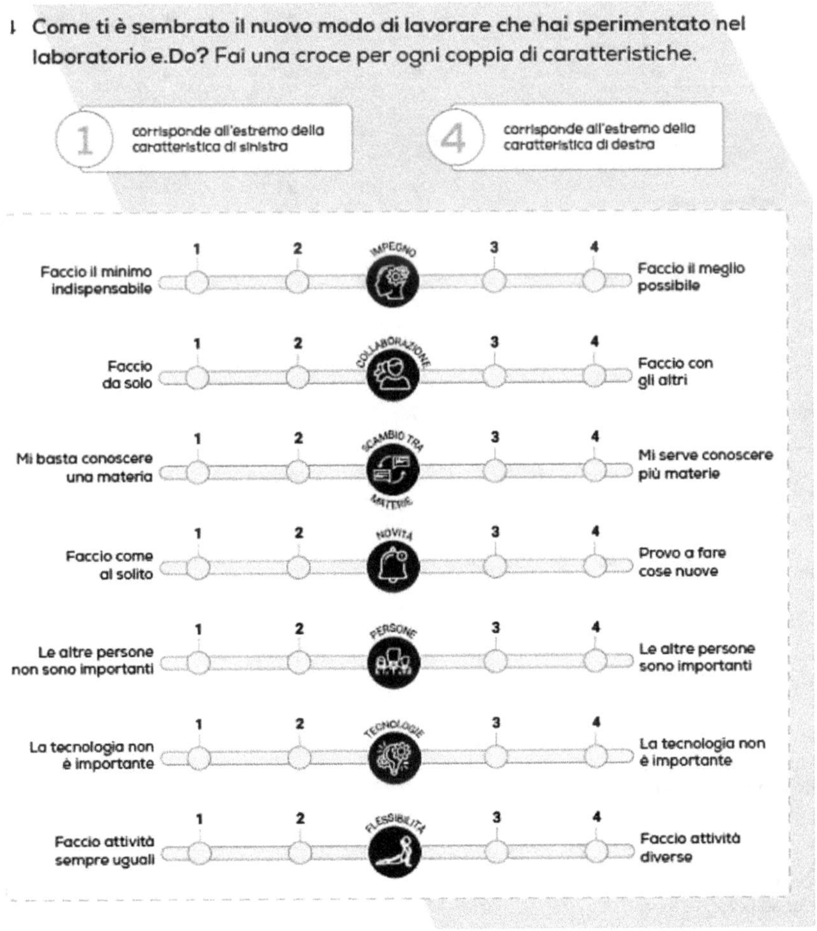

3. Thirdly, **sentences** were proposed recalling elements of facilitation of the learning process, according to the following scheme:

The solicitation concluded with some **open questions** about what the participants liked most, what advice they wanted to give to future colleagues for a fruitful participation in the experience, what suggestions they wanted to offer to the organizers in order to improve the workshop.

Figure 2. Focus group, part 2

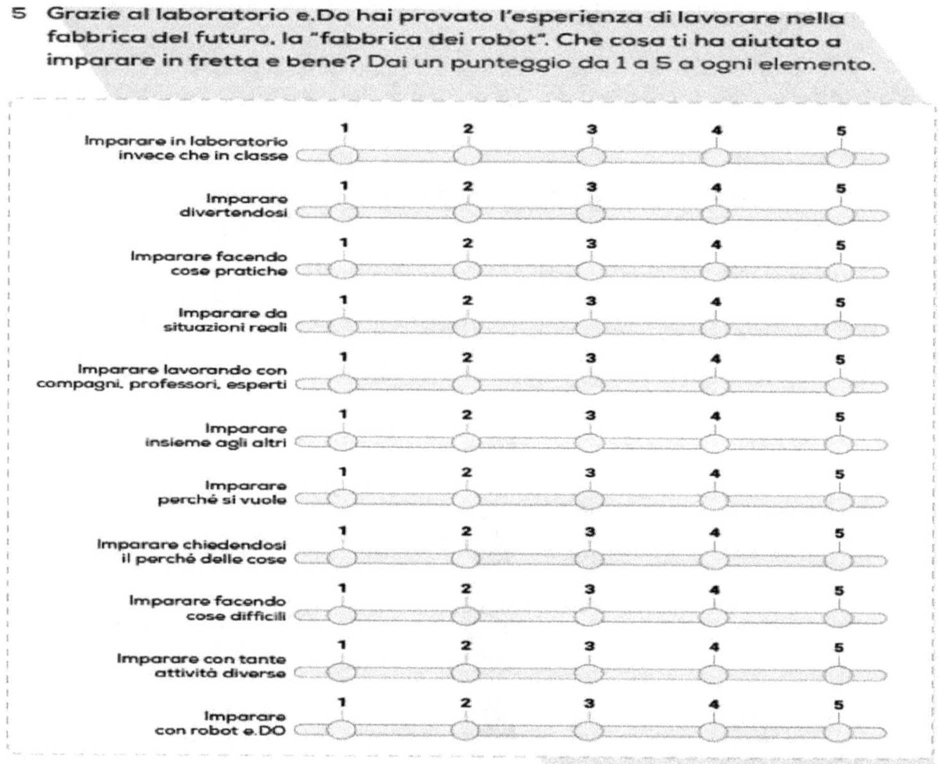

Figure 3. Focus group, part 3

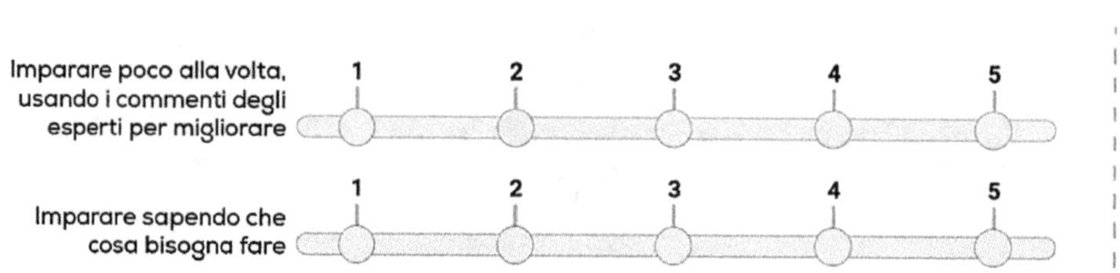

FINDINGS

The Semiotic Analysis of Images

Asking the participants to recall, through a drawing and then commenting on it, the meaning(s) attributed to the experience, generated empirical material that was as rich and interesting as it was challenging in the hermeneutic and analytical work that it required.

It was essential to rely on fundamental contributions of semiotic analysis (Ricoeur & Greimas, 2000; Greimas, 2000; Eco, 1997; Peirce, 1955) to capture the evocative provocation and the multifaceted aspects related to the density of the images, the comments and the discussion generated within the focus groups. The references of figurative semiotic analysis, in fact, solicit a functional reading both to the recognition of the image (in which conventional components are present, linked to cognitive categories/types used to attribute a meaning [Eco's *hypo-icons*, 1997], and to the narrative structure linked to dimensions of present temporality (past/present/future; sequence of phases; polyvalent frame), to different movements (incoactive, durative, terminative) and to different registers (semantic, syntactic, pragmatic). At stake is the possibility to refer to **geometric** (point, line, drawing), **visual** (shape, size, texture), **compositional** (location, direction, rhythm, symmetry) and **functional** (expressive value) elements to **grasp the value of the expressed visual communication**.

Primary School

In order to illustrate the evocative power that the collected material was able to express, let's start with **some examples** of emblematic drawings related to the focus groups involving **primary school students**.

The **first drawing** presents a compositional marking in which a series of representations, artifacts and a person connected to a dynamic movement (the arrow) are symmetrically positioned in the upper part and, in the part below, two series of letters. Having in mind the material structure of the e.DO robot, it is easy to identify the robot in the cylindrical artifact, symmetrically flanked on the right by the person (which in this case is Luca, the Comau facilitator/tutor who accompanied the workshop). As if to say that the use of the robot facilitates the person (and at the same time is in turn facilitated by the person), in reference to a pragmatic register that the sentence above

Figure 4. Primary school, example 1

conveys (the robot does the heavy lifting for the man). The artifact at the top left and the letter sequences at the bottom are not immediately traceable to conventional categories. They are clarified through the commentary the subject gives to the drawing:

"I drew a bridge which is there [on the left], then I made the tower, there is written 126 on it, which means 126 millimeters. On the bridge there is instead written 127. Then I drew Luca [the facilitator] it came out a little bad ... Then I wrote robot equals heavy work and I wrote e.DO because we played with e.DO. We used e.DO which is a beautiful robot and then I put an A+ because that is the best thing of the day."

The **metaphor of the bridge** is indicated, something that connects, that brings us closer (machine-robot and human subject), a representation of a historically dynamic and changing movement, which has characterized the entire evolution of the relationship between knowledge (scientific and empirical), technology and organizational and productive systems, from the first industrial revolution onwards. The comment also describes a **beautiful and engaging experience**, able to leave a "mark" that the student represents, referring to the world of school, with a grade: A+. The presence of the e.DO robot, therefore of the technological dimension, gives back the idea that this has captured the attention of the author of the drawing. A robot that carries out "heavy work" and therefore relieves the human being of tiring tasks. Consequently, it is possible to identify a **positive image of the**

Figure 5. Primary school, example 2

machine that assumes the function of support, in which the human being is responsible for the design phase - of a bridge, a task assigned to the students during the workshop - while the robot is entrusted with the tasks of practical implementation of the activity.

The **second drawing proposed** conveys an articulated movement, with several figures in the scene that are positioned symmetrically but also sequentially: conventional recognition highlights the children (small at the bottom) who relate to the e.DO robot (on the right) through the mediation of the explanation of the larger figure at the top (the facilitator/tutor) who illustrates the '*machinery*' of the object of common interest.

The commentary of the author of the drawing reinforces the communicative semantics conveyed:

"I drew all four of us, Luca explaining what e.DO is, and then I drew e.DO".

At stake is the **dimension of group and interaction with others, as well as with technology**, in a dynamic in which the facilitator guides participants through the workshop, illustrating the information and notions to be learned, sharing the explanation with the group, in which individuals offer their distinct and differentiated contribution. In fact, the symbols that characterize the different figures can be deciphered: for the males a ball, for the females a flower, to signify the **personal contribution** that each person, according to his or her inclinations, brings to the team.

Lower Secondary School

The figurative representations of the **lower secondary school students** also highlight interesting aspects related to the meaning of the lived experience.

A first series of drawings communicatively conveys a meaning of **reconciliation between person and robot**, that both collaborate in configuring an adequate relationship with the world. In the last figure at the bottom right, the prefiguration of a dissolving conquest of robots over the world (conventionally expressed by social representations often conveyed by literary and cinematographic tales and narratives) is opposed (*'or'*) to an alliance between the two dimensions.

Moreover, it is interesting the recognition of **emotional tensions, linked to fears, uncertainties and affective reassurances**, well expressed by the following drawings emblematically expressive of the dual tension at stake to regulate and modulate (Figure 7).

Figure 6. Lower secondary school, example 1

Figure 7. Lower secondary school, example 2

What emerges is the **link between people and technology, and the possible relationship of mutual dependence between one and the other**: on the one hand, without the human being whose task is to program, the machine is unable to carry out any operation; on the other hand, only with the help of robots the man can take a big step forward in terms of productive potential and ability to transform his world. A world that is placed at the center, the point of connection between person and robot, evoking on the one hand the image of collaboration between the two, and on the other the idea that the interaction between the two subjects - and also those that took place during the workshop - takes shape within a context that is not given once and for all, but that can be modified and improved.

The teachers' position seems to capture the tension just mentioned, as emblematically represented by the drawings shown in Figure 8.

The visual communication of the drawing on the left describes a sequential narrative in which the presence of the robot (on the left) mobilizes **interest and involvement of the students** (on the right), introducing **novelty** and generating an activation of **creativity** (the light bulb at the top right), almost playing an **inspirational function**. The image on the right evokes the extensive

Figure 8. Lower secondary school, example 3

dynamics capable of **enlarging and expanding the space of action** (the concentric circles and the non-closure of geometries), which the use of artifacts such as e.DO solicits and allows. In the face of this evolutionary potential, however, the alarming concern outlined in the central drawing stands out, in which the smiling and positive navigation of robotics is associated to the discouraging concern for a context (Italy represented on the right) in which expressions of excellence and technological quality are flanked by situations of digital weakness and backwardness. The realism highlighted seems, however, to be supported by a **positive tension**, as recalled in the following comment by one of the teachers:

"They used to teach us that if you do not know everything and you do not know it perfectly, you do not have even to try, while now you have a more open attitude, a more proactive attitude, which is often blocked by our pessimistic Italy, which is always pessimistic, always... So, these are beautiful experiences of optimism for the future, because the future is not just something that should frighten us, but it is something that should absolutely interest us, maybe it frightens us more than it does to you, so these experiences are good for us to realize that the future is also an opportunity and not just a fear."

Figure 9. The elements referred to the capabilities

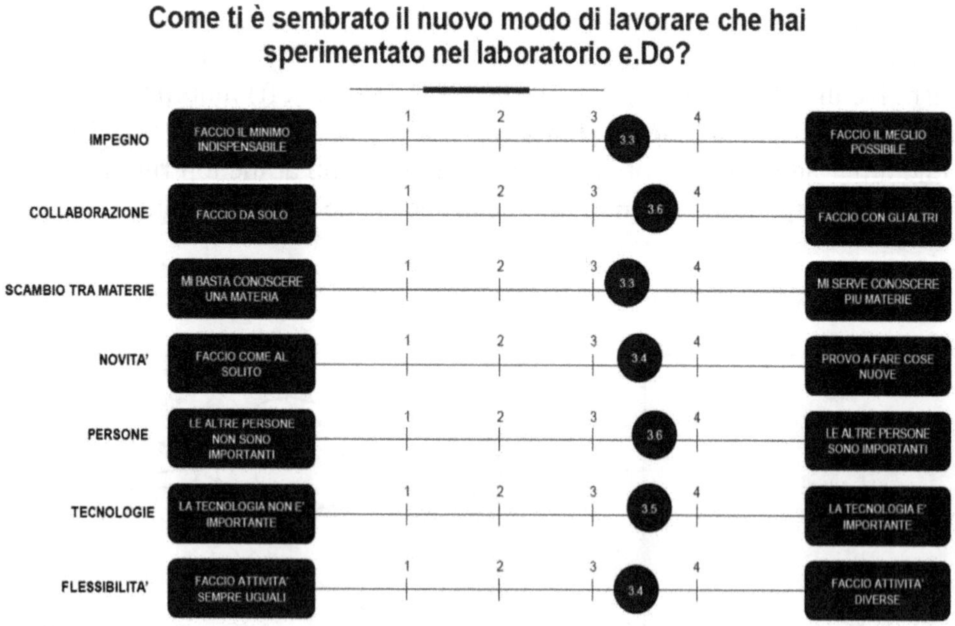

Reinforcing the **dimensions of opportunity** present in the experience, are also the positionings of the students with respect to the elements that referred to the solicited capabilities:

The relevance attributed to the **capabilities related to *Collaboration*, to *Deal with Human* and to *Deal with Technology*** is evident, further confirmed by the recognition attributed by the focus group students to the facilitating aspects found, recalled in the following figure.

The aspects of **conjunction and conjugation between fun and commitment, person and technology, individuality and group**, marking and integration of differences are all emphasized, not at all taken for granted in many traditional educational settings and essential basis for a social conception of learning processes.

Upper Secondary School

Towards a similar direction, albeit with different articulations, go the representations conveyed by the drawings collected in the focus group involving **upper secondary schools**.

Figure 10. Facilitating elements of the experience

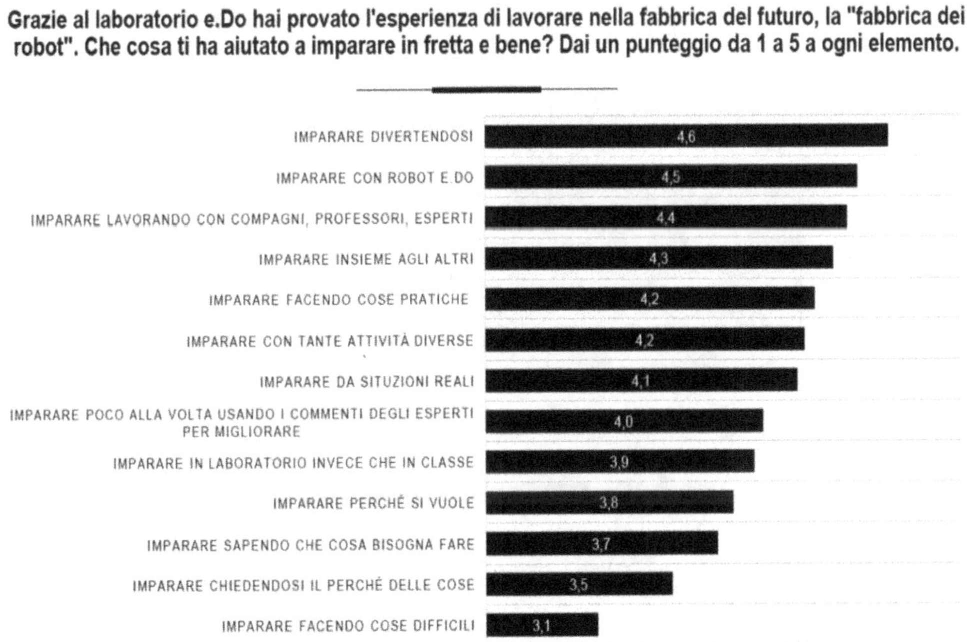

The compositional structure of the first two images in Figure 11 describes a direction and a sequentiality in which the technological artifact (the robot and its extensions) plays a part in putting the different phases in order (the numbers and arrows of the second drawing; the articulation of the technological arm that enables writing in the first drawing), producing a pragmatic outcome of curiosity, happiness, and intelligence (the words of the comic strip in the fourth drawing associated with the robot) and an evolutionary tension represented positively (the world transforming from a chaotic form to a technological configuration (the PC drawn on the top right), in the face of an observation (the person with open arms contemplating what is happening).

The **orientation to the future** evoked by the visual communications described is further confirmed by the drawing in Figure 12, in which e.DO becomes a sun beginning the dawn of a new day, with the arrow at the top pointing to the phrase *'my future'* as an emblematic evocation of a positive representation of the prospective tension at play.

Figure 11. Upper secondary school, example 1

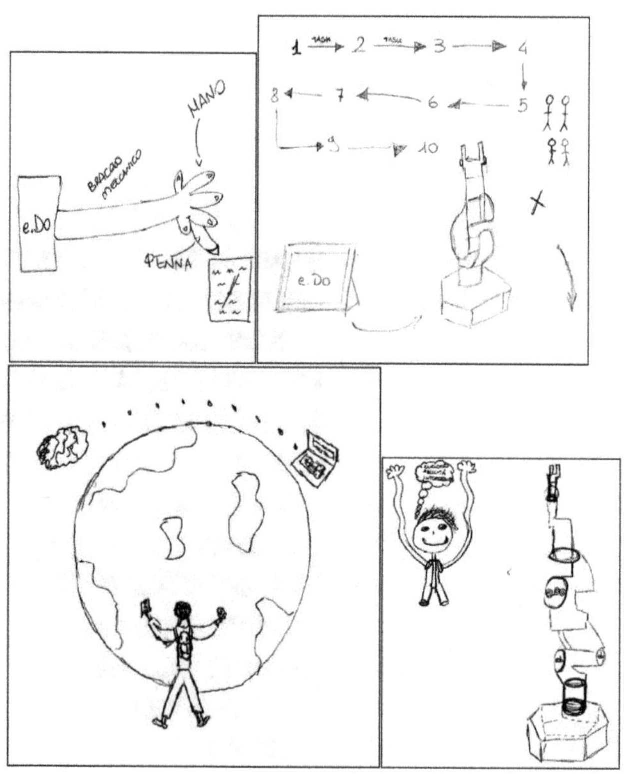

Figure 12. Upper secondary school, example 2

Teachers also seem to pick up on this compositional tension (Figure 13).

The narrative structure of the first drawing on the left describes an external gaze that takes pleasure (the illuminating light bulb) in what is happening in the laboratory, according to an articulated movement that dynamically stages the students, the robot and the tutor. In the drawing on the right, the arrow indicates a sense and a direction (from left to right and from top to bottom), expressions of a positive and evolutionary tension.

Figure 13. Upper secondary school, example 3

The teachers' verbalizations illustrate the **pleasurable amazement and generative surprise** associated with the represented experience:

"The best ally is the energy that kids can put in, when the approach to teaching is gamified...this approach will be the most productive and best lever in the near future."

"The main challenge is to have the teacher having fun... it is difficult to expect the students to have fun if the teacher is not having fun in turn, therefore, in my opinion, this is a relevant characteristic to think about."

"I attended classical high school, so I thought it was a completely different world ... however, I felt that both the kids really enjoyed this experience, I felt it was a very well structured and organized activity also for the challenges it introduces in the teaching method...".

Consistent with the evidence discussed above, are the students' placements in upper secondary focus groups with respect to the elements referring to capabilities (Figure 14).

Figure 14. The elements referred to the capabilities

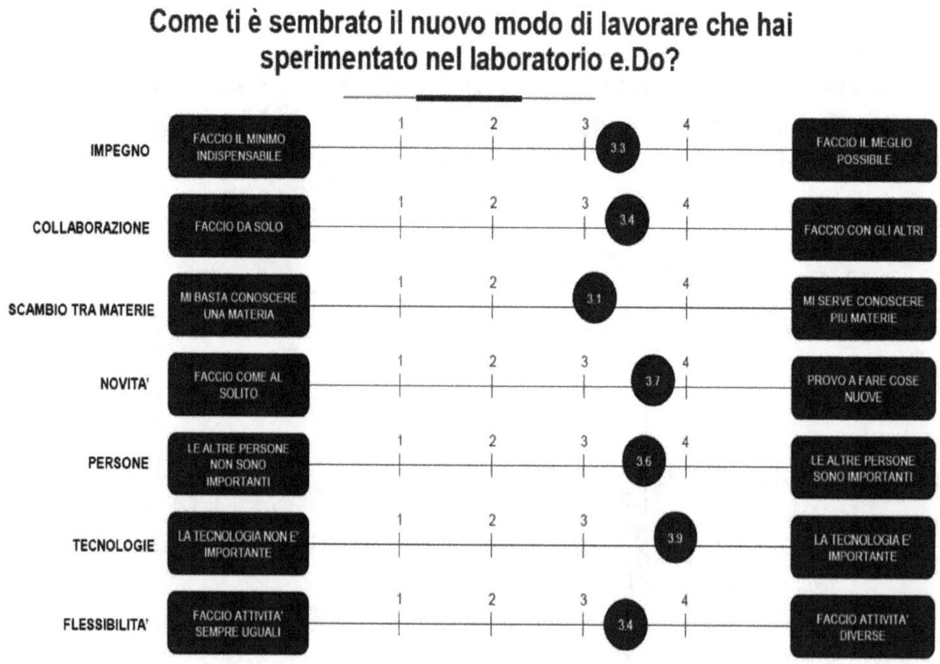

Figure 15. Facilitating elements of the experience

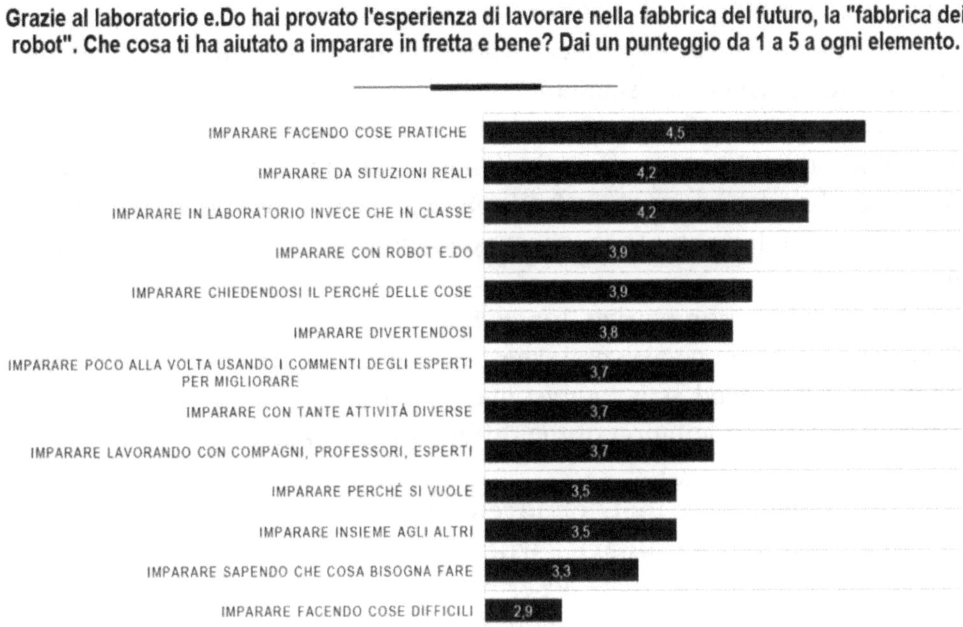

Also in this case, the ***Deal with Technology* and the *Deal with Human* are recognized as strong dimensions**, supported by an orientation to the possibility of **experiencing new things** (*Innovation*).

Even more emphasized, in this focus group, was the opportunity to access **didactic-training settings based on practical activities, on real situations, through the mediation of innovative technologies** (in this case the robot) capable of generating and stimulating **reflective skills** and **shared experimentation** (see Figure 15).

CONCLUSION

A first cross-sectional finding that emerged in the focus groups was the emphasis on having come into contact with an unexpected and curious activity that differed positively from ordinary school experiences. **Involvement, interest, development and future** are recurring thoughts within the responses provided during the focus group activity.

A second recurring element has been the prominence of the **technological contribution offered by the presence of e.DO robots**, the solicitations to face problems in groups and the choice of the treated topics perceived as an innovation with respect to traditional training paths.

A third mentioned factor relates to the **centrality of people and integrated relationships**, which are fundamental to the collaborations that can take shape facilitating the learning journey, but also expanding the chances of success in addressing complex challenges and topics. The various actors (peers, teachers, the workshop facilitator) are perceived as integral to the activity and fundamental to its success, practicing understanding, listening and collaboration.

It is possible to summarize, for each of the three school levels explored through the focus groups, a pivotal aspect around which the experience revolved, and which functioned as a catalyst and organizer of the same, feeding the attention and the solicitation of the different mobilized capabilities.

- For **primary school children**, the central element was the **relationship of familiarity with modern technological tools**. Most of the learners, in fact, described the workshop as an experience lived together with e.DO, perceived not as a machine "other" than itself, but rather as a companion, a friend with whom a relationship of complicity is established. The confidence with technology of these children, belonging to the Z Generation category, appears both deep and rooted. For this reason, they do not perceive e.DO and the tablet that controls it as elements extraneous to their context, but as playmates.
- For **lower secondary school students** (10-12 years old) the pivotal aspect is the **balance to be achieved between the human being and the machine**, in this case the robot, torn between widespread social representations of threatened machine invasion and the astonished discovery of the e.DO robot as an interesting resource that can accompany them in the process of discovery and learning, allowing them to familiarize with the world of robotics. Hence, the possibility of learning by trial and error, finding the most appropriate solution for each situation, experimenting different strategies to solve a problem until finding the one that is deemed most suitable. This "doing" is one of the key elements that is detected as a transformation of the educational dimension towards an emotional and playful dimension of learning. The words fun and curiosity are mentioned several times by the students. These aspects are accompanied by a **redefinition of the**

student's role in the learning process, as a co-protagonist and active participant in the activities of the workshop experience.

- Finally, for **upper secondary school students**, the pivotal dimension concerns **the technology used and the reference to the processes of innovation and transformation connected to it**. Students report their experience with robots in positive tones, seeing in this type of machine resources with very high potential and great capabilities, in some cases even greater than those of humans. Associated with this is the propulsive challenge associated with the requirement to solve problems by applying individual knowledge in a shared way. Management of pressure, relationship with others and organization of time become factors supporting the use of logical and reasoning skills, thought and reflection, passion, and desire to learn.
- For **teachers**, the interesting impact relates to the acquired awareness that the teacher is not the only source of learning, as well as that information and **skills are built in the interaction between multiple perspectives: school, peers and company**. The subject of learning is not only the individual student but also the group, and the presence of others provides insights, references and comparisons that help to grow.

The empirical data recalled and the dimensions described seem to support the positive impact of the workshop experience configured through the integration between the company (Comau), the university and the school, creating a course that proved to be of great interest to those who took part in it. The experiential and participatory training allowed the development of reflective, interactive, social and situated learning (Lave and Wenger, 1990), creating communities of practice and learning (Gherardi et al., 1998) among the various subjects involved. The technological dimension of the activity has played a crucial role in terms of strong impact on primary school students, as well as on secondary school ones. Generally speaking, in fact, the youngest ones are those characterized by a bond of great familiarity and affection with technology, since they were born into a society that already made extensive use of it. For them, the paradox to be faced is that of being born in a digital society and finding themselves inserted in prevalently analog school contexts. Hence, the **potential that experiences such as the *e.DO Experience* can nurture and develop** for educational quality and relevance, capable of facing the challenges of the future.

REFERENCES

Eco, U. (1997). *Segno e inferenza.* Einaudi.

Gherardi, S., Nicolini, D., & Odella, F. (1998). Apprendimento come partecipazione ad una comunità di pratiche. *Scuola democratica, 26,* 247-264.

Greimas, A. J. (2000). *Semantica strutturale: ricerca di metodo* (Vol. 2). Meltemi Editore srl.

Lave, J., & Wenger, E. (1990). *Situated learning: Legitimate peripheral participation.* Cambridge University Press.

Peirce, C. S. (1955). *Philosophical writings of Peirce* (Vol. 217). Courier Corporation.

Ricoeur, P., & Greimas, A. J. (2000). *Tra semiotica ed ermeneutica* (Vol. 4). Meltemi Editore srl.

Conclusion

Investing in Robotics Educational initiatives for the future and for the new generations means taking a long-term view and vision and taking on the challenge of imagining and building new and important learning processes. In this perspective, the issue of Value Creation becomes central for organizations (whether companies, schools and universities or institutions) and for people interested in developing new skills adapted to the challenges of the changing context in which we live and work today (Pinto et al., 2021; Lee et al., 2017; Vial 2019).

A first important finding that emerges from this research is that **the idea of Creating Shared Value is based on the interconnection of several elements**, both tangible and intangible, both theoretical and practical.

The first, more general element, which relates precisely to the theme of shared values (Porter, Kramer, 2001; 2019) is that of the **commons** (Ostrom, 1990; De Angelis, 2003; 2006;2017; De Angelis, Harvie, 2013; Centemeri, 2018), related to mutual and collective interests, to organizations understood as propositional agencies, to the possibility of developing generativity as a new way of thinking and acting (Stiegler 2010; 2010a; Magatti, Giaccardi 2017; Gherardi, Magatti, 2014; Bradley; Marcia, 1998). A second element concerns **multistakeholdership,** both internal and external (Galuppo et al., 2014; Scaratti et al., 2017), because investing in value production with a strong orientation to technological innovation and therefore to the implications of robotics use - as highlighted in the book - means mobilizing multiple internal actors, interlocutors, people, professional figures and as many external players, institutional, political, organizational and corporate, associations and communities. A third element of interconnection is what has been described as an **authentic corporate value chain**, which opens a further discourse pertaining to the theme of values, on which the configuration of early skills through digital robotics education insists and which is that of ethical responsibility, of ethical business. This has important implications for both organizational culture and for an approach to organizations and management from a critical management studies perspective (Alvesson and Willmott, 1992, 1996; Alvesson and Deetz, 2000; Alvesson and Kärreman, 2007; Alvesson et al., 2008).

A second relevant finding that emerges from the research is concerning the possibility, thanks to digital robotics education, of **promoting the development of enabling capabilities to live and work in the 4.0 world** (Fregnan et al., 2020; Ivaldi et al. 2021; Pinto et al., 2021; Warner, Wäger, 2019). This is in itself an important point of arrival that this research allows to develop, together with the opportunity to highlight how **educational processes can benefit from the mediation of robotic and virtual assistances**, which was then the specific subject of our contribution.

The combination of these two main coordinates leads to the possibility of a strategic investment in educational pathways that must be rethought as future-oriented educational experiences, capable of effectively linking **technologies to humanity**. This is a salient point that must be at the base of the educational investment. The **human factor** becomes a glue that connects and holds these aspects together.

In this possibility lies the space that this research has opened to educational initiatives focused on robotics, with a particular focus on **Educational Robotics**. The volume illustrates the Comau *e.DO Experience* project and the robot e. DO as a social agent, as a social robot (Manzi et al 2021; Baraka et al 2020;), i.e. as a tool that can be used in different contexts (Billard 1997;), consistently with the distinction highlighted between Educational Robotics and Robotic Education.

The analysis of such a robust Educational Robotics initiative – the *e.DO Experience* project – highlighted a third important finding, that is, the **value generated by educational robotics initiatives** can be expressed in terms of:

1. *Levels of value*; for **those who participate** in the initiatives, for the **companies that promote** them and for the **territory they belong** to
2. *Areas of value*; each level of value generated contains several areas of value, identified, described, and measured during the course of the research
3. *Indicators of value*; each area of value in turn can be expressed through specific indicators. These too has been identified, described and measured thanks to this research.

In fact, what emerged is a result not only for the participants in the Comau *e.DO Experience* project, but also for those who organized (the Company) and for the Territory that hosted it. In this regard, it is essential to reiterate that all this was possible thanks to the creation of an **ecosystem** made up of companies, universities, young people (Pinto et al., 2018) with a view to multistakeholdership (Galuppo et al., 2014; Scaratti et al., 2017): the initiative, created by a network of interconnected subjects, has returned value to an ecosystem.

The qualitative interviews conducted by the authors to the Comau managers and partners involved in the *e.DO Experience* design and delivery highlighted the specificity of each level, area and indicator of value examined.

With respect to the **level of "Value for Participants"**, the interviews have shown how it is possible to create value on four different areas:

- *Skill Creation*: The Comau *e.DO experience* supports the creation of different kinds of skills enabling Industry 4.0, namely "Robotic Skills", "School Subject Skills", "Professional Hard Skills". An extremely relevant point of interest is the relevance of the capabilities related to the area "Working in the Future Skill" (Pinto et al., 2021). The latter, explored in previous research (Fregnan, et al 2020) are classified in "Deal with Human", "Deal with Technology", "Agility", "Engagement", "Collaboration", "Interdisciplinarity"; "Innovation";
- *Empowerment*: The "Empowerment" area of value is the possibility to develop participants strengths and is manifested through the following indicators: "Being Satisfied and Enthusiastic for Overcoming a Challenge", "Being Aware of your own Qualities", "Being Confident and Constructive", "Being Passionate and Energetic", "Being Proactive and Enterprising";
- *Inspiring and Orienteering for Working*: this area was manifested through the indicators: "Having Galvanizing Experience"; "Getting in touch with Professional Models"; "Providing Professional Trajectiories"; "Developing Greater World of Work Awareness"; "Developing a Greater Awareness of One's Strengths and Weaknesses";
- *Value for New Citizenship*, inherent to an ethical and responsible social belonging, was manifested through the following indicators: "Challenging and Stepping out of Comfort Zone"; "Respecting Others"; "Being Exemplary and Consistent"; "Keeping Promises"; "Measuring Result Objectively"; "Cultivating Responsibility and Reciprocity"; "Pursuing Equity and Justice"; "Being Transparent and Sincere", "Being open to comparison and exchange"; "Meeting the timeframe"; "Celebrating Success".

It was also possible to decline the **"Value for Company" level** on several areas, ranging from aspects related to the development of Brand and Core Business to elements of a deeper nature. In fact, elements of true social responsibility also emerged, perceived by all the people who worked on the *e.DO Experience*: technicians, workers, managers, recognized the real relevance of this kind of initiatives, even though they are part of formal obligations for companies like Comau and are official items in the sustainability report. The *e.DO Experience*, as emerged from the interviews, also represents a possibility for organizational personnel to obtain professional recognition and to develop their skills through an exchange and continuous contamination with all the different actors encountered (whether colleagues, students, teachers, partners, etc.). The areas identified from the interviews for the level "Value for Company" can be summarized as: "Core Business/ Core Mission"; "Brand (Reputation/Marketing)"; "Employee Branding"; "Customer Social Responsibility (Formal Obligation/True Commitment to create a better World)"; "Open Innovation"; "Employee Exposure".

An interesting food for thought is given by the awareness that the type of planning required to give life to articulated initiatives with a high technological impact such as *e.DO Experience* is the same as that required to innovate and to do so in an open way: today, those who think they already have all the necessary capacity for organizational development and success in their innovation and development departments are exposing themselves to an enormous risk. Even a leading technology company like Comau knows that there is immense potential for innovation outside its own perimeter and that it resides in the new generations. Being able to discover and exploit it was truly emblematic: not only teaching, but also learning from what's out there.

The last level of value highlighted by the interviews, "**Value for Territory**", is expressed in terms of "Educationa Innovation", "Creation and Dissemination of a Culture" of education and work based on new technologies, as well as in function of the "Creation of New Ecosystems" able to support and sustain the design and implementation of the initiative and, finally, "Contribution to the Creation of Skills" and enabling capabilities to live and work in the world of work 4.0.

Subsequently, the **Survey** allowed to **measure perceptions of utility** linked to value creation on the three key levels already identified of Participants, Company and Territory.

As far as **Participants** are concerned, it was to be expected that the "Skills Creation" area would represent the one most impacted by value creation, coming to weigh more than a third of the total. Of considerable interest is also the observation that the other three areas of value generated for the participants have, in any case, a considerable weight: concretely it means that Educational Robotics has a value of 27% in "Inspiring and Orienteering" the participants towards training and professional paths linked to new technologies, "Empowerment" has a value of almost 23% and - here the most welcome note - the creation of value for the "New Citizenship" reaches almost 15%. The surprising news is that through a robot we can transmit human values and citizenship: through a robot we can be better human beings. It was also interesting to discover that the experience developed around a robot - so fascinating - is also able to inspire and orient work paths in young people who have had the opportunity to interact with it.

On the other hand, as far as the **Company** is concerned, the analysis of the data allowed to highlight how the value generated goes beyond the indicators linked to the "Core Business" (17.02%) and to the "Brand" (20.13%), elements that are important but not sufficient to exhaust the richness of these initiatives. The Educational Robotics promoted through e.DO has in fact had a relevance in terms of authentic "Corporate Social Responsibility" (13.74%), driven by ethical and generative guidelines rather than formal impositions, which weigh only 8% on the creation of Value for Company. The richness of the *e.DO Experience* also lies in its link with the openess of a company towards innovation (15.53%), capable of producing positive effects on "Employee Exposure" for its people in terms of development of skills and professional recognition (11.30%).

Finally, with regard to the **Territory**, we return to the concept of ecosystem that, through the robot, has managed to create and disseminate culture, an element that for respondents weighs 25.83% of the total value for the Territory, to innovate education, an element considered most significant with a weight of 33.02% and to contribute to the spread of the skills necessary to live and work in the twenty-first century, an aspect that weighs 26.77%. Through the experience of a common investment in robotics in an educational key, a new work culture is created to better manage the Fourth Industrial Revolution, seizing its possibilities, mitigating its risks and, above all, putting the human factor at the center of technological development. Any future initiative in this field is evident that it must be done together: a single company, no matter how large, a single university, no matter how prestigious, and individuals working individually and not interconnected cannot generate this type of value; in this regard, respondents to the Survey attribute to the dimension "Creation of New Ecosystem" a weight of 14.27% on the value for the territory.

Analysis of the works produced by **students and teachers** shows how the *e.Do Experience* leaves its mark. It is not a casual encounter; it is an extraordinary encounter. Our lives as human beings are characterized by encounters: encounters and people who have left a mark, who have made us fall in love with passions and professional paths. It clearly emerges that the encounter with the robot is an encounter of technology and the future, but also an encounter that brings us back to humanity, to being human. And this is perhaps the most comforting message that emerges from the **perception of the participants**. They have certainly grasped the essence of the new way of working, they have had experience and they recognize it, especially as regards the dimension of working together with others: the technological challenge is won by collaborating and working together. Another important fact is the recognition of educational innovation linked to learning drivers, such as the playful, engaging and experiential dimension.

On the whole, the results of this scientific research provide coordinates, tools and approaches to look in the future to the involvement of multiple stakeholders and also to the acquisition of concrete levers that can help the crossing of boundaries between entities, players, operational realities and various subjects; institutional, political, cultural and economic. With the *e.DO Experience* it has been described a challenge, a challenge that for the authors has been successful for the reasons described before, and that is declined through several verbs that they have often referred to: "to approach", "to familiarize", "to open to innovation", "to integrate". Attitudes, dispositions but also behaviors that have been nurtured, that the *e.DO Experience* has highlighted as practicable, possible and expendable. This brings us back to the human factor that has already been emphasized and recalled as a central element of this experience of value generation. It is possible to decline several challenges underlying the choice to invest in this factor, which the *e.DO Experience* has concretely highlighted.

- First, it is at the *organizational level* that it becomes a central dimension of reference, precisely in the effort not to take it for granted, but to put the person at the center even in his or her relationship with new technologies;
- There is then an implication for the institutional *educational dimension*, both scholastic and academic, where the human factor translates into attention to new educational and training needs and demands that come from people. Here again, the human factor is decisive, because we really put the students, the schoolchildren at the center of attention;

- There is a third implication that concerns the *practitioners* who are the teachers, who are the university teachers, who are the educational operators, who are required to have new professional skills: it is necessary to instill digital skills not only in the recipients, but also in the people who have to preside over the educational and training processes;
- There is also an implication linked to *families and social stakeholders*, who are then the bearers of the demand to make the development process grow in this direction: what is needed is the acquisition of an awareness of citizenship, of a truly open and evolving educational demand that holds together technologies and the human factor;
- And finally, there is a last implication for *students*: the challenge is to be urged to a more proactive approach to study, from receptive passivity to the entrepreneurial curiosity that the robot-mediated experience has amply highlighted, especially in the qualitative part of the empirical evidence we have produced.

In conclusion, we are faced with the possibility of spending this experience for a new way of value generation, towards different goals for both the production industry and the organizational social systems, looking for a finalization for sustainable and generative solutions, which go beyond a small and medium-term vision, but really look towards the future. The experience that is produced in the book says that this is not only possible, but also constitutes a strategically sustainable enterprise.

REFERENCES

Alvesson, M. (1996). Leadership studies: From procedure and abstraction to reflexivity and situation. *The Leadership Quarterly*, *7*(4), 455–485. doi:10.1016/S1048-9843(96)90002-8

Alvesson, M., & Deetz, S. (2000). Doing critical management research. *Sage (Atlanta, Ga.)*.

Alvesson, M., Hardy, C., & Harley, B. (2008). Reflecting on reflexivity: Reflexive textual practices in organization and management theory. *Journal of Management Studies*, *45*(3), 480–501. doi:10.1111/j.1467-6486.2007.00765.x

Alvesson, M., & Kärreman, D. (2007). Constructing mystery: Empirical matters in theory development. *Academy of Management Review*, *32*(4), 1265–1281. doi:10.5465/amr.2007.26586822

Alvesson, M., & Willmott, H. (1992). On the idea of emancipation in management and organization studies. *Academy of Management Review*, *17*(3), 432–464. doi:10.5465/amr.1992.4281977

Baraka, K., Alves-Oliveira, P., & Ribeiro, T. (2020). An extended framework for characterizing social robots. In *Human-Robot Interaction* (pp. 21–64). Springer. doi:10.1007/978-3-030-42307-0_2

Billard, A., & Dautenhahn, K. (1997). Grounding communication in situated, social robots. In *Proceedings Towards Intelligent Mobile Robots Conference.* Report No. UMCS-97-9-1, Department of Computer Science, Manchester University.

Bradley, C. L., & Marcia, J. E. (1998). Generativity-stagnation: A five-category model. *Journal of Personality*, *66*(1), 39–64. doi:10.1111/1467-6494.00002

Centemeri, L. (2018). Commons and the new environmentalism of everyday life. Alternative value practices and multispecies commoning in the permaculture movement. *Rassegna Italiana di Sociologia*, *59*(2), 289–314.

De Angelis, M. (2003). Reflections on alternatives, commons and communities. *The Commoner*, *6*(Winter), 1–14.

De Angelis, M. (2006). The World Social Forum: Challenging empires. *Development*, *49*(2), 125–128. doi:10.1057/palgrave.development.1100264

De Angelis, M. (2017). *Omnia sunt communia: On the commons and the transformation to postcapitalism.* Bloomsbury Publishing. doi:10.5040/9781350221611

De Angelis, M., & Harvie, D. (2013). *The commons.* Routledge.

Fregnan, E., Ivaldi, S., & Scaratti, G. (2020). HRM 4.0 and New Managerial Competences Profile: The COMAU Case. *Frontiers in Psychology*, *11*, 11. doi:10.3389/fpsyg.2020.578251 PMID:33329228

Galuppo, L., Gorli, M., Scaratti, G., & Kaneklin, C. (2014). Building social sustainability: multi-stakeholder processes and conflict management. *Social Responsibility Journal.*

Gherardi, L., & Magatti, M. (2014). *Una nuova prosperità: Quattro vie per una crescita integrale*. Feltrinelli Editore.

Ivaldi, S., Scaratti, G., & Fregnan, E. (2021). Dwelling within the fourth industrial revolution: Organizational learning for new competences, processes and work cultures. *Journal of Workplace Learning*. Advance online publication. doi:10.1108/JWL-07-2020-0127

Lee, M. X., Lee, Y. C., & Chou, C. J. (2017). Essential implications of the digital transformation in industry 4.0. *Journal of Scientific and Industrial Research*, *76*, 465–467.

Magatti, M., & Giaccardi, C. (2014). *Generativi di tutto il mondo unitevi! Manifesto per la società dei liberi*. Feltrinelli.

Manzi, F., Di Dio, C., Di Lernia, D., Rossignoli, D., Maggioni, M. A., Massaro, D., Marchetti, A., & Riva, G. (2021). Can you activate me? from robots to human brain. *Frontiers in Robotics and AI*, *8*, 14. doi:10.3389/frobt.2021.633514 PMID:33681301

Ostrom, E. (1990). *Governing the commons: The evolution of institutions for collective action*. Cambridge University Press. doi:10.1017/CBO9780511807763

Pinto, D., Scaratti, G., & Fregnan, E. (2018). *Giovani, Università e Azienda: il nuovo perimetro formativo*. Pearson.

Pinto, D., Scaratti, G., & Fregnan, E. (2021). *Learning Agility 4.0: ecosistemi e trasformazione culturale*. Pearson.

Porter, M., & Kramer, M. R. (2011). Creare valore condiviso. *Harvard Business Review Italia*, *1*(2), 68–84.

Porter, M. E., & Kramer, M. R. (2019). Creating shared value. In *Managing sustainable business* (pp. 323–346). Springer. doi:10.1007/978-94-024-1144-7_16

Scaratti, G., Ivaldi, S., & Frassy, J. (2017). Networking and knotworking practices: Work integration as situated social process. *Journal of Workplace Learning*, *29*(1), 2–23. doi:10.1108/JWL-06-2015-0043

Stiegler, B. (2010). For a new critique of political economy. *Polity*.

Stiegler, B. (2010a). *Taking care of youth and the generations*. Stanford University Press.

Vial, G. (2019). Understanding digital transformation: A review and a research agenda. *The Journal of Strategic Information Systems, 28*(2), 118–144. doi:10.1016/j.jsis.2019.01.003

Warner, K. S., & Wäger, M. (2019). Building dynamic capabilities for digital transformation: An ongoing process of strategic renewal. *Long Range Planning, 52*(3), 326–349. doi:10.1016/j.lrp.2018.12.001

About the Authors

Ezio Fregnan is Comau's Academy Director, for which he coordinates worldwide innovative training programs for executives, professionals, and young talents. He was also Comau's Human Resources Training Director, for which he managed company training processes in 12 different countries around the world. He was also member of the FCA Global Training Committee, Coordinator of the FCA Global Manufacturing Training Community, Fregnan is an expert and Executive PHD Researcher on Work Organizational Psychology, Organizational behaviors, Change Management, Train the Trainers, Instructional Design, Educational Robotics, Industry 4.0 and Digital Transformation. He is also a member of General Counsel of Confindustria Innovative & Technology Services. Affiliate Professor at Catholic University of Milan and ESCP Europe Business School of London, he is the coordinator of many specializing and executive master's degree programs, focused on Industrial Automation, Digital Transformation, and Innovation & Technology Management. Currently, he is attending an Executive Phd on Work Organization Psychology. Ezio Fregnan is author of numerous publications, including: "Gestire la formazione in una grande azienda", Guerini e Associati, 2006; "Project and People Management: an operational toolbox", Mc Graw Hill, 2011 published globally in Italian, English and Chinese; "Managing Challenges Across Cultures" Mc Graw Hill 2015 published globally in English and Chinese; "Giovani, Università e Azienda: il Nuovo Perimetro Formativo", Pearson 2018.

Donatella Pinto is co-founder and Senior Partner of Arca Partners, an Advisory Company specialised in manufacturing related activities. Donatella is also lecturer at the Catholic University of Milan of the Work Psychology and Smart Working Course within the Master of Science Innovation & Technology Management since 2018. She is the author of the books '' Learning Agility 4.0 – Ecosystems and Cultural Transformation '', "Young people, universities and companies: the new training perimeter" and "Project and People Management – An operational guide. The Comau experience in the world". She carried out her career in the Stellantis Group (formerly FCA) holding various positions in the field of human resources, dealing with various projects of Change Management, Merger & Acquisitions and Development of Strategic Plans, to support the business. In FCA she covered several roles of Human Resources Business Partner: Fidis S.p.A. Group, Consumer Services Business Unit, Alfa Romeo Business Unit and Commercial Operations area. From January 2007/ December 2020 Donatella has been Chief Human Resources at Comau (Automation Systems Company part of FCA) and since October 2019 to May 2021 held simultaneously the position of Head of Global Learning at FCA. Then, always in Comau, Donatella has been Consultancy and Education Business Unit Leader from January 2021 to December 2021.

Giuseppe Scaratti is Full Professor of Work and Organizational Psychology at the Department of Human and social sciences, University of Bergamo, Italy. His research interests include the study of knowing, learning and change in organizations; reflexivity in organizations; qualitative methodologies for the study of organizational life; evaluation and assessment of complex and transformative actions. He is Past Coordinator of the Italian Psychology Association – Section: Organizational Psychology (AIP), associated with EAWOP (European Association of Work and Organizational Psychology), from 2009 to 2013. He has taught Work and Organizational Psychology (English course in Master of Science in Management, Faculty of Economics, Catholic University of Milan), and Smart working and organizational psychology (English course in Master of Science in Innovative Technology Management, Catholic University of Milan) in the past years. Prof. Scaratti has published numerous articles and chapters, published in Management Learning, Qualitative Inquiry, Social Responsibility Journal, Journal of Workplace Learning, Qualitative Research in Organizations and Management: An International Journal, Journal of Business and Technical Communication, Journal of Health Organization and Management, Learning, Culture and Social Interaction, Journal of Management Inquiry, Human Relations, British Journal of Management. He is also a founding member of Digital KBC (Knowledge Boundaries Crossing) and of the CREIAMO (Center of Research, Education, Innovation, Action Managerial and Organizational).

Index

H

I

K

L

M

O

P

Q

R

www.ingramcontent.com/pod-product-compliance
Lightning Source LLC
LaVergne TN
LVHW081336300926
849161LV00008B/1982